FULL OF BEANS

FULL OF BEANS

PETA LYN FARWAGI

HARPER COLOPHON BOOKS

To Derek

This book was originally published by Pierrot Publishing Ltd in 1978. It is here reprinted by arrangement.

First edition HARPER COLOPHON BOOKS 1978

LCCN 77 17642
ISBN 0 06 090601 4

Printed in Great Britain by Hazell Watson & Viney Ltd
Aylesbury, Bucks, U.K.

Cover illustration by Tessa Traeger
Book design by Julian Rothenstein
Illustrations by Ian Pollock

CONTENTS

INTRODUCTION

My travels to all corners of the world as British Vogue's travel editor, coupled with a love of spicy, international foods and a curiosity about tribes and local customs, resulted in my collecting traditional recipes wherever I went. One day when sifting through this collection I realised how many of them were either based almost entirely on beans or used beans as one of the main ingredients. I also realised that what I had was a unique combination of healthy and mainly inexpensive dishes that could bring back the memories of bean feasts I had enjoyed – up the Amazon, down the Nile, in fragrant Bali seated beside a rickety old stall, in Tunisia under a black tent in the desert – as well as recapturing visions of local markets spilling multi-coloured beans on to the crowded side-walks under sunny skies. I set about recreating these happy times for my husband and friends, as closely as possible, and found I was actually giving them many dishes they had never before heard of, let alone tasted.

To a certain extent there was an element of competition involved because my husband, having grown up in the Middle East, was already one lap around the track before I had begun. For a long time I had been used to him surreptitiously slipping a piece of lemon rind into my green lentil soup or doubling the cumin in my Ful Medames; his excuse for this dastardly interference being his years of communion with the Middle Eastern bean and the pronouncements of his inviolable Alexandrian grandmother, who like Durrell's Leila in the Quartet, never went out of the house but oversaw every minute detail of the cooking for her beloved family.

Strictly speaking, my involvement with the multifarious world of the bean goes back to one Dr Ulrich Gaster Williams, my grandfather, who was a descendant of Lady Godiva and a New Zealand missionary called Henry Williams. He was a wonderful eccentric, known throughout New Zealand as U.G. (I called him Gastric Ulcer). He happily treated Maoris and Europeans alike with his patent nature cure medicines and wrote books expounding his theories on healthy eating. He passed on his enthusiastic beliefs about the importance of the role of the bean in a staple diet to his son, my father, who adapted them for his family. Thus we ate at home every variety of bean, including sprouts and peanuts, in as many different ways as my father's visionary imagination could dream up.

I have also included some recipes that are the inspirations of my remarkable mother. I give these recipes because, although they cannot be traced specifically to one country, their ingredients are exotic and their ideas have been gleaned from her various travels. I call them Mum's Bean Delights.

P.L.F.
London, 1977

WHAT BEANS ARE

Just what are beans? What are pulses? How can a pea or a peanut be a bean? And legumes – aren't they just a word in the French language for vegetables?

Well, it's all very simply really. Bean is the name given to the edible seed of the gargantuan Leguminosae family, a family of some 15,000 species, the second largest plant group in the world, that also includes some well known inedible delights such as laburnam, broom, gorse and the gorgeous wisteria. Pulse is a collective term for the edible seeds of leguminous plants – beans, peas, lentils; while legume is the edible pod that contains the seeds (such as a runner bean). Peanuts, perhaps surprisingly, are members of the Leguminosae family.

NUTRITIONAL VALUE IN YOUR DAILY DIET

As well as being delicious and bringing the tastes of the world into our kitchens at a reasonable price, beans are very healthy food. They are all high in proteins, especially the soya bean and the peanut, and are also rich sources of iron and other minerals. Soya beans should be as regularly used in meals as potatoes or rice as, weight for weight, they are as high in proteins as meat (see the bean list for ways of using them in countless natural ways) and contain fats that are effective in reducing the cholesterol level in blood.

All the beans are excellent food fibre (commonly known as roughage) – always an essential part of a sensible eating pattern. Most people do not eat enough fibrous food because so many of the products on sale today are refined, so beans sprinkled into salads, or mixed into loaves and stews keep up a steady supply and also help digestion. Miraculously for such a sensible and versatile food, beans are relatively inexpensive to buy.

BEANS IN HISTORY

The earliest known references to beans are in about 5000 BC and the direct ancestors of the ones we serve on our tables today have been found in prehistoric Peruvian graves, in Mesopotamia and in the tombs of the Pharoahs for whom they were supposed to provide on their journey in the after-life. After these auspicious beginnings they led a fairly chequered career. In Ancient Rome they were thought to be associated with death and were denounced by Pythagoras in the 6th century BC who had such a horror of them that he was killed on the edge of a bean field rather than set foot in it whilst trying to escape from the people of Crotona.

However, beans were rated by Plato as the means to a long and healthy life and were turned into a sort of Cuisine Romanique by Apicius in the world's oldest surviving cookbook. The Greeks and Romans used broad beans as voting chips—white ones were 'for' and black 'against'. In the bible too, pulses are held in esteem; Daniel would not defile himself with meat and lived with his young friends on a diet of peas while Esau sold his heritage for a mass of lentils.

Later, Columbus started a to-ing and fro-ing that was to bring Old World beans (chick peas, lentils, etc) to the New World and New World beans (kidney, haricot, etc) to the Old where they became the delicacies of the Court of Isabella and Ferdinand in Spain. Dried beans and peanuts were the staple diet of sailors on the marathon exploratory voyages and were all the nourishment offered to the slaves on the 18th and 19th century slavers. In Europe, peas and beans became very popular during Medieval times. Desert Arabs still carry lentils by the camel-bag full on their nomadic wanderings.

Today, almost every race has its own national dish based on beans. It is almost always a rich, full-flavoured recipe that arises out of a combination of the local history, the traditional ways of cooking and the availability of beans, meats and herbs.

Preparing and cooking beans

1: When and where to buy

People tend to think that, as beans are dried, they can be bought at any time and anywhere. This is not true. The older they are the longer it takes to cook them and the more likely it is that they will go soft on the outside and stay hard in the middle. Beans of bad quality and those which have not been dried quickly will never achieve the plumpness nor the individual flavours that you expect from them. So, as their prices do not vary by more than a penny or two from shop to shop, buy them at a good delicatessen or health food store where you know the turnover is quick and where you can see what you are buying. That done, they should last for 6-9 months in your own cupboards . After that time they start to lose flavour and take longer to cook. The rarer varieties, such as flageolets, come into the shops in the Autumn and have to be bought in tins for large parts of the year. In fact, tinned beans although much more expensive are enormously useful for quickie meals for hungry children or friends who drop in unexpectedly, as they can be heated in a few minutes on top of the stove or put into a casserole with other ingredients and be ready in 30-45 minutes. Personally I do not recommend tinned bean sprouts because they are soggy and wilted-looking with none of the fresh, crispy bite of the ones bought by weight in supermarkets, Chinese stores or sprouted in your own home. Whether you are baking or using raw peanuts in recipes, it is a sensible idea to buy them in large quantities because it is much cheaper, they do not deteriorate quickly and because of their high nutritive value it is a good plan to have them readily available at all times – remember that a handful of peanuts and a glass of milk make a full-protein snack for all members of the family.

2: Storage

Keep the beans, peas, lentils and peanuts in a dry, cool place, preferably in a glass or metal jar. Their warm colours are an asset to any kitchen and kept in glass jars they look good on an open shelf or on a work surface well away from the warmth of your cooker or stove.

3: Soaking

It will not do any harm to any of the beans or peas to be soaked overnight (except peanuts of course) although varieties such as lentils can be cooked straight from the packet. For the exact soaking times see the list of beans following.

There are two ways of soaking beans. The traditional overnight or 8–12 hours which has been the way since prehistoric times or the quicker method of bringing them to the boil, holding them there for 2 minutes and then letting them soak for 2-3 hours. The latter is extremely useful if you discover half-way through the day that you wish to serve beans for supper or dinner.

In either case there are several points to remember:–

1: Pick out any beans that are an odd colour and sift through for grits. Even the best bean will have one or two in them.

2: Soak beans in plenty of cold, clean water in the ration of 2:1 of water:beans

3: After the first 15 minutes or so, once the skins have begun to wrinkle and stretch, pick out any beans that rise to the top.

4: Do not soak longer than the suggested times because the beans will start to get sticky.

5: Keep the water the beans have soaked in and use it for cooking because some of the vitamins will be wasted if it is thrown away. This does not apply for certain dishes where the colour of the water would spoil the dish.

6: Remember, when experimenting with your own bean recipes that dry beans, once soaked, approximately double their weight and volume.

4: Cooking

I feel very strongly that the best way to cook beans is long and slowly. This way they absorb more of the spices and seasonings and will cook to tender plumpness. A rushed bean is not a very happy one. See the list following for the timings for each variety, remembering that the times given are approximate because they depend on the quality and age of the bean. This is a good time to explain why I have not given pressure-cooking times and here are the reasons:

(a) it is virtually impossible to gauge to the minute how long each different kind of bean will take and it is not very practical to keep cooling down and opening the pressure cooker to check.

(b) the fast cooking tends to produce beans which are crumbly around the edges and hard inside.

(c) the whole point of a bean casserole or stew is the long slow absorption of spices and herbs.

(d) it is only possible to cook a small amount of beans because you have to allow room for swelling and space for the vacuum.

(e) some beans foam more than others and chick peas, for example, produce scum which has to be skimmed off; either the scum or bubbles could interfere with the valve on the top of the pressure cooker.

(f) the short preparation time and long, slow cooking means you spend less time in the kitchen and can get on with other things rather than spend your time hovering over a pressure cooker.

The basic suggestions for cooking are:

1: For cooking as vegetables boil the beans, for the times suggested, in clear, cold water or their own soaking water. The water left may be used at a later stage of the recipe or kept for stocks or in soups if the colour allows.

2: For stews and casseroles, cook the beans in the same water until reduced to the required amount. This will vary from recipe to recipe and will also depend on the colour of the finished dish. Instructions are given in the recipes.

3: Do not use baking soda in the beans as it destroys some of the vitamins.

4: Do not add salt at the beginning as it will stop the beans from swelling and plumping out. Add it about 15-20 minutes before the end of the cooking time.

5: Some skimming may be necessary, particularly with chick peas and this should be done after 20 minutes if they are cooking on their own and as indicated in the recipe if there are other ingredients involved.

5: Utensils

The utensils you use depend on how you are cooking your beans. The one common utensil is a wooden spoon which is always used so that the beans don't get damaged and broken. The pans are obviously very important and a short moment's reflection before starting the recipe will make a great difference. Where important, I have mentioned the kind of pan in the recipes. As a general guideline any *large* pan will do for boiling beans on their own as a vegetable because the boiling, however slow, will keep them circulating in the pan. For dishes prepared on the top of the stove, on burners, heavy cast iron is best because the heat will be evenly distributed; a lighter pan will overcook the beans on the bottom and they will go soft before the rest are done. For dishes that go into the oven, use either cast iron, earthenware or anything else that is ovenproof. It is not necessary to use a bean pot with its small opening at the top although it does make a change from a regular casserole or straight-sided English pot and sets the dish apart.

When cooking recipes with bean shoots it is certainly easiest to use a Chinese wok which is a metal pan shaped like a bowl for quick, effective heat distribution. However these recipes I have given can all be cooked in a frying pan or skillet.

6: How to Freeze

It may seem an unnecesaary idea to freeze beans because you think they are always ready and waiting on your shelf. I find it very useful to use the freezer for the stews and casseroles which I can make up in large amounts and keep in packets in the freezer as they only take a short time to defrost. I almost always keep some kind of bean and meat stew with basic seasoning only in the freezer because I can whip it out and turn it into a curry, chilli con carne or something similar by just adding the right spices and seasonings. A little more liquid will probally have to be added when heating. Beans can be kept in the freezer for 5 months.

Spilling the beans

ADUKI BEANS

An excellent, very tender bean with a unique taste. These small, shiny, red beans from China and Japan go into casseroles, soups, loaves, bean and sweet rice dishes and also into desserts, breads and muffins. Delicious with a mixture of chopped mushrooms, onions, tomatoes, rice and basil to stuff aubergines and peppers. Mix with zuccini for an unusual salad.

Soak for 2 or 3 hours and allow 1-1½ hours to cook.

BLACK BEANS

These sweet, glistening ebony beans are popular in Spain and in Central and South America. They are delicious in casseroles and soups and particularly good flavourings with them are cumin, garlic, bayleaves and tomatoes.

The longer they are soaked and cooked the better, so soak overnight and boil them slowly for 2 hours (they will cook in 1-1½ hours if the heat is increased).

BLACK-EYED BEANS

A whitish bean/pea with a black or yellow eye which was originally brought into America by the slaves from Africa. They are still very much bound up in tradition ; many people eat black-eyed beans and rice on New Year's Day to ensure good luck in the coming year and these beans also feature in voodoo ceremonies.

The Californian ones are big and plump and the Turkish ones, which are more often found in Europe, are slightly smaller. They are very popular in the southern states of America.

Soak overnight and cook for 1-1½ hours.

BORLOTTI BEANS

Used often in Italian cooking, these light brown/pinkish beans with their wine coloured speckles are particularly good in dishes or soups calling for nutmeg, mace or cinnamon because the beans become very soft and absorb these aromatic spices (and herbs as well). When cooking in a casserole, try them with fresh Parmesan cheese sprinkled generously on top before serving. These beans are also sometimes available in a white form.

Soak overnight and cook for 1–1½ hours.

BROAD BEANS

A very versatile bean in either its fresh or dry state. Although found in Switzerland in the Bronze age, it is a Mediterranean bean which the Greeks and Romans favoured and also used for voting; the white was 'for' and the black 'against.' In the Middle Ages the Britons cooked broad beans with everything.

When fresh they should be eaten as young and green as possible, on their own, or in purées, pies and salads. Dried, they vary in colour from creamy-white to warm brown and respond well to long cooking in rich meat stews.

Soak overnight and cook for 1½-2 hours.

BUTTER BEANS

Often known as the Madagascar bean, because many are grown there, the butter bean, as its usual name implies, is a pale, unsalted butter colour. It is large, flat and kidney-shaped. Cook in milk, purée with cream with a knob of butter on top and sometimes a little Parmesan cheese. Serve with roast pork for Sunday lunch. Butter beans are a favourite in the English kitchen.

Soak overnight and cook for 1-1½ hours.

CHICK PEAS

A less familiar bean, nutty of flavour with a fine, crunchy consistency. They have long been a part of the staple diet of Spain where they are known as (garbanzos), Italy (ceci), India (chana dal), the Middle East and parts of France.

Toasted and salted, they make good partners for drinks; puréed and mixed with sesame oil and seasonings they make Hummus bi Tahini which is a highly seasoned dip. Italians and Spaniards stew them with garlic and rough peasant sausages, often with the unusual addition of nutmeg; and they can add unexpected nuggets of flavour to a spinach salad or an attractive topping to a rice dish. Street sellers in Egypt sell them ground with wheat, seasoned and deep fried as Felafel.

Columbus brought them to Europe on his second voyage to the West Indies. They taste rather like mild chestnuts and look like hazlenuts and vary in colour from the creamy beige European and American ones to the dark brown Indian ones that are sensational with fruits in curries.

Soak overnight and cook for 1½-2 hours, skimming the top after about 20 minutes.

CHINESE BLACK BEANS

Smaller than the South American black bean, about the size of a haricot, they have a green layer under the skin and are not as sweet in flavour. The Chinese love them fermented in salt or stewed and served with soft fish or crab.

Soak overnight and cook for 1½-2 hours.

DUTCH BROWN BEANS

These are an especially good tasting variety of the haricot bean that go into those gorgeous plump-looking Dutch bean-pots. But remember they do require shelling and allow the considerable time needed.

Soak overnight and cook for 1½-2 hours.

FLAGEOLETS

These rare, gorgeous pale green beans are a little longer and thinner than the regular kidney bean shape, and have a very pleasant, delicate taste. They are grown in France and Italy and often are only available elsewhere in tins. But they are worth tracking down – and can usually be found in Italian delicatessens in their dried form in the autumn. They are very good with roast lamb when they should be cooked until only just tender.

Soak overnight and cook for 1-1½ hours.

FUL MEDAMES

Called either Ful or Foul or sometimes even Fool, the names are somewhat misleading because Ful in Egyptian only means bean and the Medames explains that they are a special kind of broad bean, smaller and plumper than ours and of a light coffee colour. however *all* Egyptians adore them and are constantly stopping in at a Ful café for a bowl and so the generic name Ful has become a nickname. There are many ways of eating them but the most habitual is stewed with cumin then covered with olive oil, and garnished with hard-boiled eggs, raw onion and a peasant salad at the side (see page 153).

Soak overnight and cook for 2-2½ hours.

HARICOT BEANS

A small, plump, tender white bean that is one of the main ingredients in many classic casserole dishes all over the world. It can also be served as a purée or in salads. There's a lovely, cosy smell like bread baking when they are being cooked.

Soak overnight and cook for 1-1½ hours.

GREEN OR SNAP BEANS

Fresh beans that only need about 4 minutes cooking in boiling salted water. Use them in marinaded salads and add to stews and casseroles about 20 minutes before completion. The fresher and younger you buy them the healthier, tastier and crunchier they will be — or even better is to grow them yourself. They are easy to cultivate in almost any spare patch of land and respond well to warmth and water.

GREEN LENTILS

Lentils come in various sizes and colours varying from the red split lentils through the small, plump, purple-green ones from China to the large brown or green lentils. Because they are eaten all over the world there are endless recipes from such diverse places as Austria and the Seychelles. There are many delicious variations of the lentil or lentil and rice soup. They are also delicious served as a strongly seasoned purée with ham, pork or sausages.

Lentils need no soaking and the small ones cook in about 20-30 minutes, the larger ones about 30-45 minutes.

MUNG BEANS

These are one of the best beans to grow bean sprouts from because of their sweet flavour (see page 144). The smallest of the regularly eaten beans, they are olive-green coloured and squarish in shape. Originally only used in India and South East Asia, but since the wide interest in health they have found the place on our tables that their exceptionally high vitamin content deserves. Mung beans can also be used as a purèed vegetable, in soup or in a stew with pork.

Soak overnight and cook for 1 hour.

LIMA BEANS

These pale green lovelies abound in vitamin C. They are better known to the American cook than to the Europeans who usually have to buy them tinned. The smaller creamy-white ones are known as butter limas; both are kidney shaped. Use them in salads, loaves, as delicious stuffings for red or green peppers and most famous of all, for Succotash (see page (72) – an old American dish of limas, corn, molasses and paprika. Delightful.

Soak overnight and cook for 1-1½ hours.

NAVY BEANS

These are the American version of the French white haricots, the ubiquitous Beans Meanz variety sometimes known as Michigan beans. Much used in America as an all purpose bean but although we import 80,000 tons a year for the tinned market they are not usually available in our shops in the dry state. Navy beans are at their best in casseroles and stews where they have time to absorb the seasonings and can be used in any of the recipes that call for the various white beans.

Soak overnight and cook for 1-1½ hours.

PEANUTS

The surprise bean, the nut that's not a nut. Peanuts have a high protein content, second only to the soya bean and are rich in vitamin B and minerals. Sometimes known as groundnuts they are natives to America. Found in the tombs of the ancient Inca Indians they were later carried by the slave ships in the 18th and 19th centuries for the obvious reason that they were cheap and nourishing and practical to store. Peanut butter was made centuries ago in South America by grinding the nuts and mixing with honey. Better to make it yourself today (see page 45) than buy it because the commercial product is full of additives. Roast your own peanuts in the oven with a little oil and sea salt, or with cumin, chilli or curry powder, leaving on the skins for nutrition and fibre. Make them into soufflés, loaves and spicy chicken dishes with cinnamon or nutmeg; sprinkle them whole or ground into salads. The possibilities are endless.

PEAS, FRESH AND DRIED

Legend has it that the pea comes from the Garden of Eden and archaeologists have found them with the mummies in Egyptian tombs – presumably providing health food for the after-life journey.

Personally I like to eat peas straight from the garden, on their own cooked with fresh mint, or in casseroles such as those dreamed up by the Italians. Never use bicarbonate of soda to keep their colour as it also destroys their food value and digestibility. A treat I can never resist when they are available is the tiny piselli. Dried peas produce those rather British mushy peas with tough skins, however long they are cooked and I do not recommend them.

PIGEON OR GUNGA PEAS

These are small round flat peas, beige with chestnut-brown speckles and marks. They are used all through the Caribbean where from one island to another they form the staple diet together with rice. They can be used as a vegetable served on their own with a large knob of butter, on in rice dishes and casseroles. They are available dried and tinned in West Indian and Oriental stores.

Soak overnight and cook for 45 minutes – 1 hour.

SNOW PEAS

Flat, wide, pale green fresh peas, eaten pods and all. They should be eaten crunchy and barely cooked as the Chinese do. Either blanch them or quickly stir-fry them for a few minutes. The Western equivalent is the mange tout.

PINK BEANS

These are a variety of kidney beans, the bean of frijoles refritos (refried beans) and south western American dishes.

Soak overnight and cook for 1-1½ hours.

PINTO BEANS

Originally used only in the spicy chilli-based recipes from Mexico and thereabouts, pinto beans have now become very popular in America and are finding their way into the health stores in Europe. Particularly good with meat stew where their distinctive salmon colour speckled with brown can show off to advantage.

Soak overnight and cook for about 1-1½ hours.

SOYA BEANS

The most nutritious of all beans. Weight for weight with steak, soya beans have more proteins. For thousands of years the soya bean has been the cow of the Orient, as the milk from the bean was the only kind they drank. I am against the use of soya beans as a direct substitute for meat. Used in their own right whole in salads or loaves; as a vegetable interestingly spiced or cooked with tomatoes, garlic and thyme; purėed; sprouted; as bean curd; in conjunction with meats; as an oil; as milk in cooking other vegetables – they will always be original. The granules used as substitutes tend to be a let down because whatever you do they will never be meat. It's far better to enjoy them naturally and experiment. They are good with duck, chicken, pork, tomatoes and fish, and can be used in casserole dishes using other beans as well if the soya beans are cooked for 1 hour before adding the other beans.

Soak overnight and cook for 3½-4 hours.

BEAN CURD Fresh, custard-like cakes of pressed, puréed soya beans. Sold by the cake which is usually ½ inch thick and 3 inches square, in Oriental provision stores. Drain and store in fresh water in a covered jar in the refrigerator for up to 2 weeks, changing water daily.

FERMENTED BLACK BEANS You can ferment black beans but it usually is the black soya which is strongly flavoured and preserved. If unavailable substitute extra soya sauce or salt.

SOYA BEAN SPROUTS These are longer than the mung bean sprouts – 1½-2 inches long – and are thin, crispy, chewy and delicious. They are sold fresh by weight and in tins, the fresh ones having parchment like husks that should be removed before eating.

SOYA OIL Contains lecithin, the one fat found to be effective in reducing cholesterol levels in blood.

WHITE BEANS

Elusive to pin down and describe because they occur in several variations. The majority of these medium-sized white kidney-shaped beans come from Argentina but the Italian *Canellini* and the French *Soisson* are two tasty, fluffy specialities worth searching for.

Soak overnight and cook for 1-1½ hours.

SPLIT PEAS

The fresh yellow and bright green split peas are the most heart-warming of all, for those cold winter nights when you find the idea of cold hard to banish. The green ones are good with some chopped fresh mint, the yellow look and taste gorgeous puréed into a thick soup with cumin, coriander or just with boiled onion and a little cream.

Soak 3-4 hours and cook for 45 minutes to 1 hour.

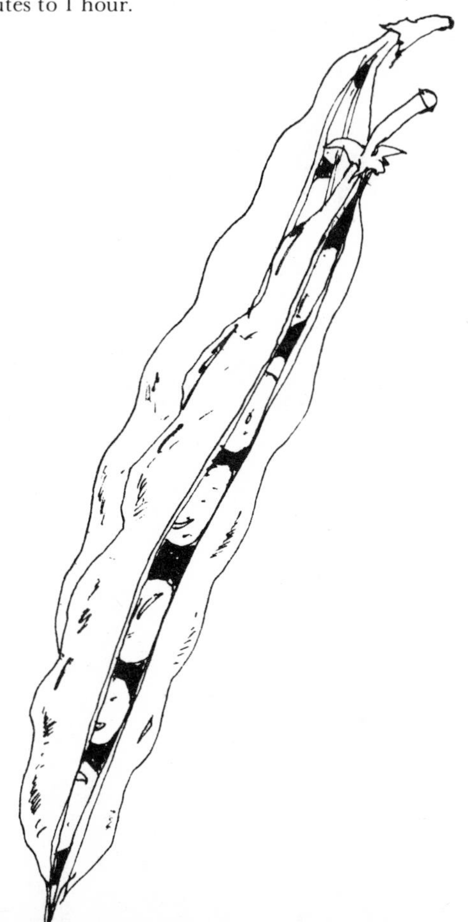

ELZEKARIA *BASQUE SOUP*

FRANCE

A strong peasant soup from the fiercely proud French Basques.

1 large onion, sliced
1 oz (2 tablespoons) pork fat or lard
1 small white cabbage, shredded
8 oz (1 cup) haricot beans (soaked, see page 14)
2 cloves garlic, peeled and crushed
freshly ground black pepper
4 pints (10 cups) water
salt to taste
a dash of vinegar

Sauté the onion in the fat in a large saucepan. Add the cabbage, beans, garlic and pepper. Pour in the water, cover the pan and simmer gently for about 3 hours until the beans are tender. Add salt and vinegar to taste about 10 minutes before serving. Serve with rough country bread.

4 servings as a supper dish, 6 as a starter

SERBIAN BEAN SOUP

AUSTRIA

A Serbian peasant soup with the delicate flavour of juniper berries. It makes a delicious meal if followed by cold meats, bread and cheese.

12 oz (2 cups) white beans (soaked, see page 17)
½ lb smoked bacon
4 leeks, thinly sliced
2 bayleaves
5 juniper berries
2 cloves
3 pints (7½ cups) water
2 oz (¼ cup) butter
1 tablespoon flour
salt and pepper to taste
½ teaspoon marjoram
1 teaspoon herb vinegar
5 tablespoons cream

Cook the beans in an iron pot with the bacon, leeks, bayleaves, juniper berries and cloves in the water for 1½ hours. Take out the bacon and cut it into cubes and mash half the beans. Melt the butter, add the flour and fry until it bubbles, then add the whole beans, mashed beans, bacon, salt, pepper, marjoram, vinegar and the cooking liquid from the beans. Simmer for 10 minutes. Stir in 4 tablespoons of the cream. Serve immediately with the remaining spoonful of cream floating on the soup.

4 servings.

CHICK PEAS WITH LAMB SOUP

IRAN

Chick peas and lamb have been a favourite food with Persian nomads for centuries and were put into soups when camp was struck in the hills or by water holes where water would be plentiful. The soup is clear, light and spicy.

2 lb lean lamb
1 tablespoon butter
4 pints (10 cups) boiling water
1 onion, chopped
½ teaspoon paprika
½ teaspoon cinnamon
salt and pepper
1 lb (2¼ cups) cooked chick peas (see page 13)

In a large saucepan, fry the meat lightly on all sides in the butter to seal in the juices. Pour the boiling water over it. Cover and cook slowly on top of the stove for 30 minutes. Take the meat out and cut into chinky pieces. Return to the pan, adding the onion, paprika, cinnamon, salt, pepper and drained chick peas. Bring to the boil and then simmer for 2½ hours.

Persians eat this with very flat thin bread and the nearest equivalent outside Iran is the softer flat bread you can buy in Cypriot stores, or else you can make your own (see page *157*).

6 servings.

HUNGARIAN BEAN SOUP

HUNGARY

Popular all over Hungary, as much in Budapest restaurants as at country tables, this soup is made of three of the most used ingredients in Hungarian cooking, beans, sour cream and paprika.

6 oz (1 cup) white haricot beans (soaked, see page 14)
4 pints (10 cups) stock
8 oz (1½ cups) carrots, chopped
2 parsnips, chopped
4 oz (1 cup) celery, chopped
8 oz (1 cup) smoked ham or hungarian salami, chopped in small cubes
1 tablespoon olive oil
1 small onion, chopped
1 clove garlic, peeled and crushed
1 oz (¼ cup) flour
salt and pepper to taste
1 teaspoon paprika
4 tablespoons sour cream

Cook the beans in the stock for 1½ hours. Add the carrots, parsnips, celery and smoked ham or salami and simmer until tender (about 1 hour). Heat the oil in another pan and fry the onion and garlic until browned, stir in the flour. Blend this mixture into the soup, stir and boil for 10 minutes. Season with salt, pepper and paprika and mix in the sour cream just before serving.

Serve in deep soup bowls with dark brown or black bread.

4 servings.

BOHENSUPPE *BEAN SOUP*

AUSTRIA

In Austria they celebrate a Bohenkonigsfest (feast of the bean's king) January 6th – Epiphany. He who finds the bean which has been baked into a cake is the king of beans. This thick bean soup can make a simple feast on many an icy winter's night.

3 oz (½ cup) white beans (soaked, *see page 17**)*
2 large onions, 1 chopped – 1 sliced
1 oz (2 tablespoons) fat
3 tablespoons flour
3 oz (½ cup) diced cooked ham
2 pints (5 cups) stock or water

Cook the beans until soft. Fry the chopped onion in half the fat until golden, add the flour and when the onion and the flour are browned, add the mixture to the beans and gradually stir in the stock. Simmer for 10 minutes. Let cool and purée in a blender until smooth. Put the soup back into the pan, add the diced ham and cook gently until the ham is heated through. In the meantime fry the second onion (sliced into rings) in the rest of the fat.

Serve the soup piping hot, garnished with the fried onion rings.

4 servings.

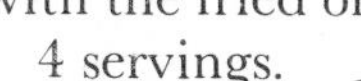

ZUPPA DE CECI *CHICK PEA SOUP*

ITALY

A favourite recipe in the Dolomites especially in winter, where the rough mountain salami provides a delicious foil for the smooth nut-flavoured chick pea. Although a soup, it is substantial enough for a whole meal.

3 tablespoons olive oil
2 medium-sized onions, chopped
1 green pepper, seeded and chopped
1 clove garlic, crushed
a pinch of red pepper flakes
1 bayleaf
1 lb (2 cups) chick peas (soaked, see page 13)
1½ pints (4 cups) chicken or beef broth
10 thin slices salami
salt and papper
a bowl of freshly grated Parmesan cheese

Heat the oil in a large heavy pot, add the onions and the green pepper and cook over low heat until softened. Add the garlic, a pinch of red pepper flakes, the bayleaf and, 1 minute later, the drained chick peas and chicken or beef broth. Simmer slowly until the chick peas are tender, about 1½ hours. Remove about ½ cup of the peas, force them through a sieve of purée in an electric blender and return the purée to the soup. Season with salt and pepper to taste. Chop the salami into small pieces, add to the soup and simmer for about 10 minutes more.

Serve with plenty of fresh grated Parmesan and toasted Italian bread.

4 servings.

BROWN LENTIL SOUP

TRINIDAD

Cooked for me by the mother of a Trinidadian friend who lives in a white clapboard house all grown over with flowers. The nutmeg is what gives the soup its Caribbean aroma.

1 oz (2 tablespoons) dripping
1 onion, chopped
12 oz (2 cups) lentils (soaked for 1 hour)
2 pints (5 cups) water
salt and pepper to taste
½ teaspoon nutmeg
½ pint (1¼ cups) milk

Melt the dripping in a saucepan and gently cook the onion until brown. Add the lentils and water and cook until tender (about 45 minutes). Rub the lentils through a sieve with a wooden spoon. Put them back in the pan, add the seasonings and milk and simmer gently for 10 minutes. Serve with 1-inch square croûtons (made by frying bread quickly in a minimum of oil or butter).

4 servings.

GUNGA PEA SOUP

JAMAICA

This is the Jamaican version of a Caribbean soup which I have particularly liked since a childhood visit to that weird and wonderful island.

The coconut milk, which makes it so special to me, acts rather like yoghurt with its light fresh taste, and makes it very suitable as a summer food.

8 oz (1 cup) gunga peas (soaked, see page 16)
½ lb salt beef, chopped
1 onion, chopped
2½ pints (6¼ cups) water
8 oz (1 cup) corn kernels
1 pint (2½ cups) coconut milk
salt and pepper to taste

Put the gunga peas, salt beef and onion into a large saucepan. Add the water, cover and simmer for about 2½ hours until the peas are tender. Add the corn and cook for a further 20 minutes. Add the coconut milk and season to taste. Reheat thoroughly and serve with triangles of buttered toast.

4 servings.

RED PEA SOUP

JAMAICA

There are several standard ways of making red pea soup, one of which includes a quantity of pigs' tails. However, I like the gorgeous looking, roughly puréed version first cooked for me in my own home by a Jamaican girlfriend.

¼ lb salt pork, chopped
1 large onion, chopped
1 hot red pepper, chopped
1 lb (1½ cups) red beans (soaked)
2½ pints (6¼ cups) meat stock or water
½ teaspoon thyme
4 stalks of celery, chopped
a handful of fresh parsley, chopped
salt and pepper to taste
10 chives, chopped

Put the salt pork, onion and red pepper into a large iron pot with the beans and stock or water. Cover and simmer for 2 hours. Add the thyme, celery and parsley and simmer for a further 30 minutes. Allow to cool and mix in the blender only long enough to make a rough purée. Return to the pot and season with salt and pepper to taste.

Sprinkle chopped chives on top and serve with melba toast.

4 servings.

MUNG BEAN SOUP

PHILLIPINES

This soup was often cooked for my family in our home in London by a Filipino girl, called Rosita, who looked after my then tiny daughter with so much love and gentle affection.

8 oz (1 cup) mung beans (soaked, see page 15)
2 pints (5 cups) water
1 rasher of bacon, chopped
1 tablespoon soya or unflavoured vegetable oil
1 tablespoon flour
1 tablespoon soya flour
salt and pepper to taste
½ teaspoon Marmite or yeast extract
½ teaspoon sugar
1 sprig of mint, finely chopped

Cook the mung beans for 1 hour in the water. Fry the bacon in the oil in a frying pan. Mix together the flour, soya flour, salt and pepper and sprinkle over the bacon in the frying pan. Dissolve the Marmite in 2 tablespoons of the water and pour into the pan, stirring all the time until a smooth paste is obtained. Pour this into the saucepan with the mung beans and water, bring to the boil and simmer, partly covered, for 1 hour. Add the sugar and mint 5 minutes before serving. Serve with fried bread.

4 servings.

BEANCURD & SEAFOOD SOUP

CHINA

The Chinese have always enjoyed eating out of doors. The Empress Tziu-hsi, the Dragon Empress, loved to sit in the gardens of the Imperial Palace watching her favourite entertainers while nibbling absently at an array of cooked dishes which, to us, would resemble a banquet.

These days it is the parking lots of Singapore that offer the same kind of gustatory experience. When the cars have gone home at night, the stalls come out and enthusiasts go from stall to stall discussing, choosing and sampling as many as 10-15 dishes. Soups are popular both as a first course (substantial ones) and just before dessert (clear ones). This is a starter.

2 dried mushrooms (from Chinese stores)
2 oz (¼ cup) thinly sliced pork
a pinch of salt
1 tablespoon oil
6 oz (1 fillet) sole, or similar white fish
1 teaspoon dry sherry
1¼ pints (3 cups) water
a small bunch of watercress
4 oz bean curd
4 slices ginger

Soak the mushrooms in warm water for 2 hours. Drain, discard the stems and slice. While the mushrooms are soaking, marinade the sliced pork with a little salt. Heat the oil in a frying pan and fry the fish until golden. Place in a deep saucepan. Sprinkle the sherry over the fish and add the water. Bring to the boil, then add the watercress, bean curd, ginger, sliced mushrooms and pork. Add salt to taste.

Serve in small deep bowls with prawn (or shrimp) crackers (also available from Chinese stores).

4 servings.

SPLIT PEA WINTER SOUP

AUSTRIA

Very often this is served for early supper in the small villages high in the Austrian Alps.

8 oz (1 cup) split peas (soaked, see page 17)
4 pints (10 cups) water or stock
2 sticks celery, diced
2 large potatoes, diced
2 medium-sized onions, sliced
2 tablespoons chopped parsley
2 oz ($\frac{1}{4}$ cup) fat
1$\frac{1}{2}$ tablespoons flour
salt and pepper to taste

Cook the peas in water until soft (about 1 hour), rub through a sieve and put back in their water. Stew the celery, potatoes, onions and parsley in the fat until soft. Sprinkle on the flour, stir well and add to the pea soup. Season with salt and pepper and simmer for 15 minutes. Serve with country-style bread.

4 servings.

LENTIL SOUP WITH LAMB

MIDDLE EAST

At the time of Jesus' entry into Jerusalem on the donkey this would have been the staple diet of believers and non-believers alike. All over the Middle East today women still shop around the markets to look for the reddest lentil they can find.

1½ lb lean stewing lamb, cut into 1-inch pieces
3 medium-sized onions, 1 chopped – 2 sliced
1 carrot, chopped
1 bayleaf
4 pints (10 cups) water
1 lb (2 cups) lentils
4 oz (¾ cup) boiled rice or boiled vermicelli
salt to taste

Boil the lamb with the chopped onion, carrot and bayleaf for 2 hours in the water, removing the scum as it rises.

Strain the broth into a clean saucepan. Add the meat and lentils and simmer for 20-30 minutes. Add the rice or vermicelli and boil for 10 minutes. Fry the two sliced onions until brown and use to garnish. Serve with croûtons.

6 servings.

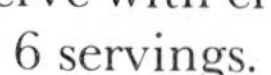

ZUPPA DI PISELLI ALL PAESANA *PEASANT PEA SOUP*

ITALY

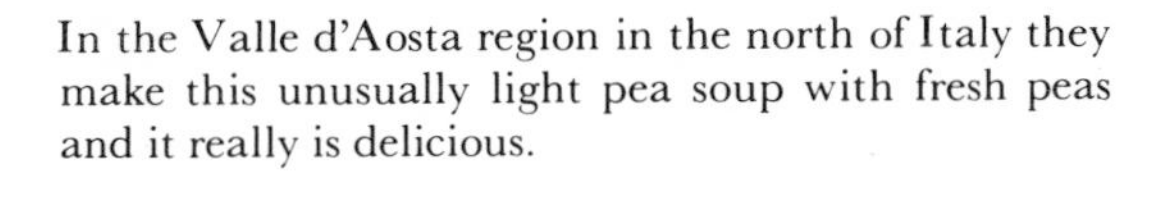

In the Valle d'Aosta region in the north of Italy they make this unusually light pea soup with fresh peas and it really is delicious.

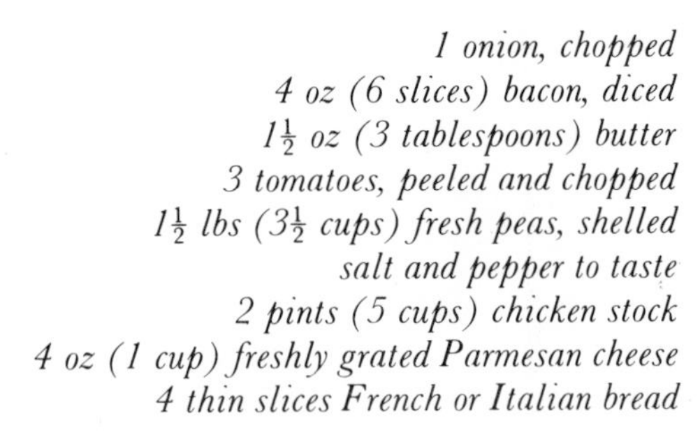

1 onion, chopped
4 oz (6 slices) bacon, diced
1½ oz (3 tablespoons) butter
3 tomatoes, peeled and chopped
1½ lbs (3½ cups) fresh peas, shelled
salt and pepper to taste
2 pints (5 cups) chicken stock
4 oz (1 cup) freshly grated Parmesan cheese
4 thin slices French or Italian bread

Sauté the onion and bacon in butter in a saucepan until the onions are lightly browned. Add the tomatoes and peas to the pan, seasoning with salt and pepper. Cook gently for 2 to 3 minutes and add the chicken stock gradually. Simmer until the peas are cooked.

Toast the bread and put a piece in the bottom of 4 soup bowls. Sprinkle the cheese over the toast and pour the piping hot soup into the bowls. Serve immediately.

4 servings.

HUNGARIAN LENTIL SOUP

HUNGARY

This lentil soup is often served as a bland starter before a spicy main course containing paprika, a traditional condiment to Hungarian cookery.

8 oz (1 cup) brown lentils
1 bayleaf
3 pints (7½ cups) water
½ onion, finely chopped
1 tablespoon olive oil
1½ oz (¼ cup) flour
2 teaspoons vinegar
1 teaspoon sugar
salt to taste
1 teaspoon mustard
4 tablespoons sour cream

Cook the lentils with the bayleaf in the water until the lentils break down (about 30 minutes). Sauté the onion in the hot oil until brown, mix in the flour quickly and stir into the lentils. Flavour with vinegar, sugar and salt to taste. Add the mustard to the sour cream and stir in just before serving. Serve with black bread.

4 servings.

SOUPE DE LENTILLES

SEYCHELLES

Seychellois cooking is greatly influenced by Creole traditions and the abundance of fresh fish; with Indian, Chinese, French and English influences also important. In spite of, or perhaps because of, all this exotica, lentil soup has remained a favourite.

8 oz (1 cup) brown lentils
3 pints (7½ cups) water
2 oz (3 slices) bacon
1 onion, chopped
1½ oz (3 tablespoons) lard
salt and pepper to taste
croûtons

Put the lentils and bacon into a large saucepan. Add the water and simmer, uncovered, for 30 minutes. Purée in an electric blender. Melt the lard in the saucepan and sauté the onion, stirring frequently, until browned. Add the lentil purée, season with salt and pepper and simmer for 15 minutes. Serve with croûtons.

4 servings.

COLD BEANS WITH OIL

SYRIA

2 medium-sized onions, sliced
2 c loves garlic, peeled and chopped
4 tablespoons olive oil
12 oz (2 cups) cooked white beans (see page 17)
1 small can tomatoes, drained
$\frac{1}{2}$ teaspoon oregano
$\frac{1}{4}$ teaspoon coriander
salt and pepper to taste

Fry the onions and garlic in the oil until brown. Add the beans, tomatoes, oregano, coriander, salt and pepper and simmer for 10 minutes. Serve cold with roast or grilled lamb.

The same recipe cooked with butter should be served as a hot dish.

4 servings.

WHITE BEAN SALAD

AUSTRIA

8 oz ($1\frac{1}{3}$ cups) cooked white beans (see page 17)
1 medium-sized onion, sliced and separated into rings
vinaigrette dressing (see page 152)

Combine the beans and onions, taking care not to break them up. Pour the vinaigrette dressing over 30 minutes before serving.

4 servings.

SOUR CREAM BEANS

2 medium-sized onions, chopped
8 oz (2 cups) mushrooms, chopped
1 oz (2 tablespoons) butter
1 teaspoon flour
12 oz (2 cups) cooked kidney beans
a scant $\frac{1}{2}$ pint (1 cup) sour cream
2 tablespoons dry sherry (optional)

Sauté the onions and mushrooms in the butter until the onions are soft and transparent. Sprinkle in the flour and stir. Add the kidney beans, sour cream and dry sherry. Simmer, stiring occasionally until the sauce thickens.

4 servings.

CHICK PEAS WITH BREAD & YOGHURT

LEBANON

This is another Middle Eastern recipe which is as old as history, and possibly older because chick peas have been found in Neolithic graves.

The cold yoghurt, fresh mint and light crunchy chick peas make it the perfect dish for a hot summer's day in the garden.

12 oz (1¾ cups) chick peas (soaked, see page 13)
4 flat Arab breads (see page 157)
1 pint (2½ cups) yoghurt, strained
2 oz (½ cup) pine nuts
2 cloves garlic, peeled and crushed
a sprig of fresh mint, finely chopped
salt to taste
a pinch of cayenne pepper

Cook the chick peas in plenty of water for 1½ hours, skimming as necessary. Drain, and retain the liquid.

Toast the bread in the oven without letting it change colour. Break it into pieces and spread out on a big open dish. Pour a little of the chick pea liquid on to the bread, then spread the chick peas over. Mix the strained yoghurt with the pine nuts, garlic, mint and salt. Pour over the chick peas and serve garnished with the cayenne pepper.

4-6 servings.

BEAN PUREE WITH YOGHURT

GREECE

The ancient Greeks, of course, used no forks and this gave me the idea that bean purée with yoghurt would make unusual picnic food.

8 oz (1⅓ cups) cooked haricot beans (see page 14)
4 tablespoons olive oil
1 tablespoon vinegar
salt and pepper to taste
2 shallots, chopped
a sprig of fresh dill (or ½ teaspoon dried)

Purée the chilled beans, oil, vinegar, salt and pepper in a blender until smooth. Serve in a shallow dish, garnished with the shallots and dill.

Serve with an accompanying bowl of yoghurt and flat Greek bread (see page *157*). To eat, tear off pieces of the bread and scoop the purée and yoghurt from the bowls.

To thicken the purée for picnics, use only 2 tablespoons oil and add 2 tablespoons yoghurt when blending.

4 servings.

See colour picture

FELAFEL

EGYPT

Felafel are to my husband's Cairo childhood what potato crisps were to mine. On the frequent afternoons when he was playing truant, the first stop would be one of the innumerable street sellers to buy a newspaper cone full of these spicy gold-brown chick pea and wheat balls. Felafel are an essential part of the Cairenes day-to-day living.

I usually double or triple these amounts, otherwise they are all gone in the kitchen before I can serve them with drinks to friends.

2 oz (½ cup) Arab bread, crumbled
2 oz (¾ cup) crushed wheat
6 oz (1 cup) cooked chick peas
2 tablespoon fresh lemon juice
1-2 cloves garlic, peeled and crushed
½ teaspoon crushed red peppers
¾ teaspoon ground cumin
½ teaspoon ground coriander
salt and pepper to taste
oil for deep frying

Soak the bread and wheat separately for 30 minutes. Drain the bread and squeeze out the moisture. Drain the wheat through a sieve. Purée the chick peas in a blender with the lemon juice, garlic, red pepper, cumin, coriander, salt and pepper, until smooth. Combine the bread, wheat and chick pea purée thoroughly and mould the mixture into little balls. Let them stand for an hour before deep frying in hot oil for 2-3 minutes. Roll them over kitchen paper to absorb any extra oil and serve hot.

4 servings.

HUMMUS BI TAHINI

CHICK PEAS IN SESAME OIL

LEBANON

Popular throughout Ancient Greece and Rome, hummus has been adapted over the centuries and now varies from country to country. I am giving the Lebanese recipe because it is the richest.

8 oz (1 cup) chick peas (soaked, see page 13)
1 or 2 cloves garlic, well mashed in salt
5 oz (1 cup) tahini (a paste of sesame seeds and oil, bought ready-made)
juice of 2 lemons
chopped parsley
ground paprika
olive oil

Wash the soaked chick peas, cook in water over strong heat until a foam appears, skim, reduce the heat and simmer for 2 hours. Purée the chick peas and garlic in an electric blender, adding sufficient water to make a smooth creamy paste. Remove from the blender and gradually add the tahini and lemon juice. Adjust seasoning to taste, adding more salt, lemon juice or garlic as necessary. Serve on shallow plates, garnished with parsley and paprika and just before serving, pour pure olive oil over the top.

In Arab countries this is usually served as part of a *mezze* (hors d'oeuvre) which includes bowls of freshly sliced tomatoes, slivers of raw carrots, gherkins, black olives,

labne (a soft cream cheese made from yoghurt, see page *154*), Arab bread (see page *157*) or thin toast.

4 servings.

BROAD BEAN AND LENTIL PUREE

EGYPT

Egyptians are lusty eaters, who for some reason I can never fathom, are always worried that they haven't eaten enough. One way they try to combat this daily problem is to whip up tasty dips which they serve with olives, raw carrots, onions, strained yoghurt and bread in the hope that when the main dish comes, their appetites will be whetted and they will be able to do it justice.

8 oz (1⅓ cups) dried broad beans
2 tablespoons red lentils
2½ pints (6¼ cups) water
1 tablespoon fresh lemon juice
salt to taste
4 tablespoons olive oil
a handful of chopped parsley.

Cook the broad beans and lentils over a low heat in the water for 1½ hours or until the beans are very soft. Leave to cool and the water left will be absorbed. Mash the broad beans and lentils in a bowl with a fork and mix in the lemon, salt and most of the oil. Chill. Spread on to a plate, pour over the rest of the oil and serve garnished with the parsley.

AKKRA *HOT BEAN PATTIES*

JAMAICA

Jamaicans need no excuse for a party – feeling good is a good enough reason – and once started the fun goes on all night. These black-eye patties come out at intervals to help keep the flagging body up to the level of the soul.

1 lb (2 cups) black-eyed beans or soya beans (soaked, see page 12)
6 fresh hot red peppers, seeded and finely chopped
salt to taste
fat for frying

Drain the beans. Rub off the skins. Cook for about 2 hours, or until plump. Beat in a mortar until smooth (or grind bit by bit in an electric blender). Turn into a deep bowl. Add the peppers to the beans with the salt and rub with a wooden spoon against the side of the bowl until the mixture becomes fluffy and almost doubles its bulk. Drop by the spoonful into hot fat and fry until golden brown. Serve for parties or before dinner drinks.

Makes 24.

PEANUT AND AVOCADO DIP

AUSTRALIA

2 oz ($\frac{1}{2}$ cup) ground peanuts
1 ripe avocado, peeled and mashed
$\frac{1}{2}$ teaspoon very finely grated onion
1 tablespoon lemon juice
$\frac{1}{2}$ tablespoon vegetable oil
salt and pepper to taste

Purée all the ingredients in an electric blender, adjust seasoning and blend again. Serve chilled with Matzos.

BLUE CHEESE AND PEANUT SPREAD

AMERICA

4 oz (1 cup) ground salted peanuts
1 oz ($\frac{1}{4}$ cup) blue cheese
1 tablespoon grated mild cheddar
$\frac{1}{4}$ teaspoon tabasco sauce
$2\frac{1}{2}$ tablespoons mayonnaise

Blend all the ingredients in a blender. Serve as a spread on bread or toast, or as a dip for crisp vegetables such as celery sticks, carrot sticks and cauliflower florets.

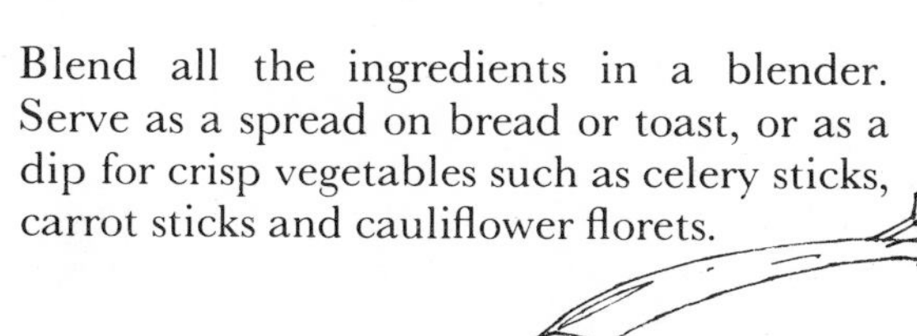

PEANUTS AND APPLE DIP

AMERICA

3 oz ($\frac{3}{4}$ cup) ground unsalted peanuts
3 oz (1 cup) grated apple
1 oz (2 tablespoons) cream cheese
$\frac{1}{2}$ teaspoon celery salt

Purée all the ingredients in a blender. Serve with raw vegetables such as carrots and celery. This is also a delicious stuffing for dates if 1 tablespoon of cream is added to the mixture.

PEANUT & CASHEW DIP/SPREAD

An unusual alternative to bowls of nuts to serve with drinks.

3 oz ($\frac{1}{2}$ cup) peanuts, raw
3 oz ($\frac{1}{2}$ cup) cashews, raw
3-4 sticks celery
2 sprigs of parsley, chopped
carrot juice

Purée the peanuts, cashews, celery and parsley together in a blender with enough carrot juice to make a thick paste. This can be used to fill celery sticks. Add a little more carrot juice for a lighter mix to use as a dip for celery sticks and crisps.

peanuts
salt (preferably sea salt)
oil

Grind the nuts in a blender, add enough oil to moisten, and salt to taste. Add more oil if necessary, until you get the desired consistency.

PEANUTS-GO-LIGHTLY DIP

A delicious dip devised by my mother.

8 oz (2 cups) salted peanuts, finely chopped or coarsely blended
4 fl oz (½ cup) thick yoghurt
¼ teaspoon finely grated lemon rind

Mix all the ingredients together and serve with cheese crackers and crisps.

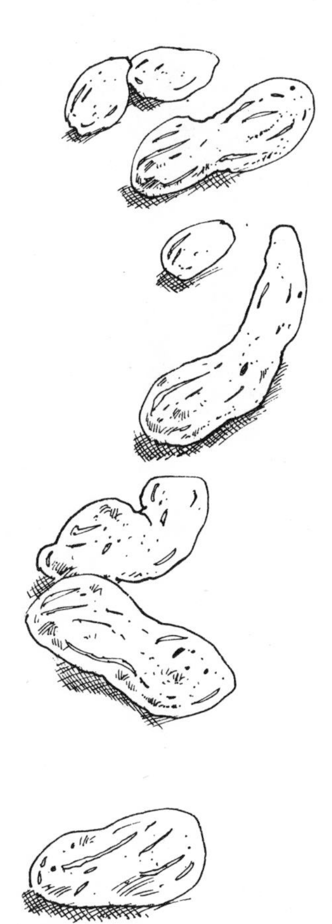

DANDELION & BEAN SALAD

GREECE

A fairly rare salad from the northern part of Greece. The dandelions must be young.

4 oz (1 cup) small dandelion buds
6 oz (1 cup) cooked white beans
2 tablespoons olive oil
1 tablespoon lemon juice

Wash the dandelions well and boil for 15 minutes. Drain and cool. Mix with white beans in a dish and pour over the oil and lemon juice. Serve as a side salad or as a first course.

4 servings.

GARBANZO NUTS

GREECE

7 oz (1 cup) chick peas (garbanzos)
salt

Soak the garbanzos overnight, drain and leave to dry. Spread out on a baking sheet, sprinkle liberally with salt and bake in the oven at 350°F, mark 4 for about 45 minutes. These can be made in large quantities as they keep well in screw-topped jars.

FRIED BEAN CURD

CHINA

A favourite Chinese snack often offered at the food stalls in Singapore parking lots (see introduction to Beancurd and Seafood soup, page 29).

oil for deep fat frying
10 squares bean curd (from Chinese stores)
1½ tablespoons soya sauce
2 teaspoons salt mixed with freshly ground black pepper

Heat the oil until almost smoking. Drain the bean curd and place in a wire strainer. Gently lower a few pieces at a time into the hot oil. Allow to cook quickly to a light golden colour then remove carefully and drain on absorbent paper. Sprinkle over the soya sauce and the salt and pepper.

Serve with any Chinese food, as a snack to eat with dinner drinks, or give to children as a healthy snack.

FEIJOADA *BLACK BEAN CASSEROLE*

BRAZIL

Feijoada is almost as sacred to the Brazilians as religion on Sundays. Even the stylish Cariocas (people from Rio) seek out restaurants serving this powerful earthy dish.

1 lb (1½ cups) black beans (soaked, see page 12)
¼ lb smoked tongue, cubed
½ lb salt beef, cubed
½ lb pork sausage, sliced
1 large onion, chopped
2-4 cloves garlic, peeled and crushed
a generous amount of fresh parsley, chopped
2 tablespoons oil
salt to taste

Cook the beans for 1 hour in fresh water. Meanwhile boil the smoked tongue and salt beef for 30 minutes. Sauté the chopped onion, garlic and parsley in the oil until the onion is soft and transparent. Add the sausage and fry for 10 minutes. Add this mixture to the beans, together with the other meats and cook, covered, in the oven at 300°F, mark 2 (or simmer gently on top of the stove) for another 2 hours.

Salt to taste towards the end of the cooking time. Just before serving, remove ½ cup of the beans, mash them well and return to the pot to thicken the liquid. Serve with plenty of boiled rice and sliced oranges.

6 servings.

PETJEL AJAM *COCONUT CHICKEN*

INDONESIA

In Indonesia half the fun of eating is in eating out, stopping at the ramshackle wood stalls with rickety benches and discussing your meal or snack with the stall holder who will cook it in front of you. Peanuts are an essential ingredient, raw, roasted or used as oil. Combined with coconut or coconut milk they provide the basis for many recipes.

1 chicken (3–3½ lb)
1½ tablespoons vegetable oil
2 tablespoons desiccated (shredded) coconut
14 oz (2 cups) rice
2 medium-sized onions, chopped
4 cloves garlic
2 dried red chillies
4-6 oz (½ – ¾ cup) roasted peanuts
1 tablespoon lemon juice
1 tablespoon brown sugar
1 teaspoon ground ginger
salt

Cut the chicken into joints and dry with paper towels. Heat the oil in a large saucepan and brown the joints all over. Take out the chicken and most of the oil, leaving only a tablespoonful in the pan. Put 2 cups of boiling water over the desiccated coconut, leave for 2 minutes and strain.

Start the rice by boiling fast for 10 minutes and then steam for a further 10 minutes with the pan covered. Fry the onion, garlic and chillies in the oil until the onion browns. Return the chicken to the pan with the peanuts and the remaining ingredients. Pour in the liquid drained from the coconut and simmer for about 40 minutes.

Arrange the rice in a ring on a large platter with the chicken pieces in the middle and pour the sauce over.

4-6 servings.

HAITIAN RICE & BEANS

HAITI

Voodoo food, from the land of naive paintings, often offered to the spirits during rites and then taken home and eaten afterwards.

7 oz (1 cup) rice
1 onion, sliced
1 large clove garlic, peeled and finely chopped
1½ tablespoons oil
a pinch of dried red chillies
8 peppercorns
1 coconut
6 oz (1 cup) cooked kidney beans

Cook the rice as usual. Meanwhile, sauté the onion and garlic in the oil until slightly brown. Add the chillies and peppercorns and remove from the heat. Retain the milk from the coconut and grate about ¼ of the coconut flesh. Add the coconut milk to the onion mixture and heat. Add the kidney beans and the cooked rice. Stir gently so as not to break up the kidney beans. Add most of the grated coconut, retaining 2 tablespoons to sprinkle on the top. Serve piping hot with a tomato and onion salad.

4 servings.

See colour picture facing page 57.

RUSSIAN BEANS AND NUTS

RUSSIA

A dish for cold Russian winters, a time when nuts and beans are stored in jars on the shelves of most kitchens.

1 large onion, sliced
2-3 oz ($\frac{1}{4}$ cup) butter
4 oz (1 cup) chopped nuts
4 oz ($\frac{2}{3}$ cup) cooked butter beans (see page 13)
4 oz ($\frac{2}{3}$ cup) cooked haricot beans (see page 14)
2 oz ($\frac{1}{3}$ cup) cooked soya beans (see page 16)
6 tablespoons ($\frac{1}{2}$ cup) sour cream
a pinch of sugar
salt and pepper to taste

Sautè the onion in the butter until golden. Add the nuts and beans, cook until heated, then stir in the sour cream and sugar and season generously with salt and pepper. Serve as a first course with sliced buttered brown bread or as a main course with boiled noodles.

4 servings.

MARINATED PORK & BEAN SPROUTS

CHINA

Pork is highly regarded by the Chinese as a special delicacy and this dish will only be cooked if a good enough quality of meat can be afforded.

½ lb lean pork
1 egg white
1 teaspoon sugar
2 teaspoons soy sauce
¼ chicken stock cube, crumbled
1 teaspoon cornflour (cornstarch)
2 tablespoons peanut oil
8 oz (4 cups) bean sprouts
2 cloves garlic, peeled and crushed
4 spring onions, finely chopped
salt and pepper to taste

Shred the pork finely. Beat the egg white until stiff, fold in the sugar, soy sauce, stock cube and cornflour and pour over the pork. Leave to marinate for at least 1 hour.

Heat the oil in a wok or frying pan until very hot and stir-fry the bean sprouts, garlic and spring onions for 30 seconds. Remove and drain. Pour off most of the oil and put the pork in the pan together with the marinade mixture. Season with salt and pepper and stir-fry for 3-4 minutes. Stir in the bean sprouts, garlic and spring onions and heat all through. Serve immediately with boiled rice in individual bowls.

2 servings.

COCIDO

SPAIN

A national dish that varies according to pocket and preference all over Spain.

8 oz (1 cup) chick peas (soaked, see page 13)
1 lb salted beef
½ lb smoked ham
1 lb chorizos, thickly sliced
1 small chicken, cut in serving pieces
1 large onion, chopped
1 clove garlic, peeled and crushed
2 teaspoons Spanish paprika
1 lb spinach
2 tomatoes, peeled and chopped
2 carrots, thickly sliced
2 leeks, chopped

Put the chick peas, salted beef, smoked ham, chorizos, chicken, onion, garlic and paprika in a big heavy pot, cover with water and bring to the boil. Skim the top and simmer, covered, for 1 hour.

Add the spinach, tomatoes, carrots and leeks and cook for 1 hour, adding water as necessary, so that there is spare broth left at the end of the cooking time.

To serve, take out the vegetables and meat, arrange the vegetables in the middle of a platter with the meat and sausages around them and strain the remaining broth into a tureen. The broth is served with Spanish or French bread first, followed by the meats and vegetables.

6-8 servings.

AUSTRALIAN HOT POT

AUSTRALIA

A dish out of the billycan from the Australian outback. As far as I know my Australian grandmother never actually saw the outback, being a gentle lady from Melbourne, but she could play the digereedoo, and she taught me how to throw a boomerang when I was four. She cooked this stew for us on visits to England on days when she refused to admit that there could be a country with such unpleasant weather.

2 large onions, thinly sliced
6 stalks celery, cut into 2-inch pieces
1 large can tomatoes, drained
8 oz (1 cup) haricot beans (soaked, see page 14)
salt, pepper and sugar
2 oz (⅓ cup) chopped raisins
6 chump chops
seasoned flour
1 teaspoon vinegar
4 oz (1 cup) grated Cheddar cheese
4 oz (2 cups) fresh breadcrumbs
1½ oz (3 tablespoons) butter

This is a layered dish and it is important to put the ingredients in the right order or the beans will not cook.

Into an oiled casserole put layers of about half of each the onions, celery, drained tomatoes and beans. Season with salt, pepper and a little sugar, and add half the raisins.

Shake the chops in a bag of flour seasoned with salt and pepper, and place in the casserole. Add the rest of the beans and

Key to the beans illustrated overleaf

1 Brown lentils
2 Green split peas
3 Kabli chan (Chickpeas)
4 Red beans
5 White beans
6 Soya beans
7 Borlotti beans
8 Continental lentils
9 Small lentils
10 Chinese red beans
11 Butter beans
12 Chinese lentils
13 Yellow split peas
14 Chinese black beans
15 Small lentils
16 Mung beans
17 Cannellini beans
18 Indian chick peas
19 Pinto beans
20 Aduki beans
21 Flageolets
22 Broad beans
23 Soissons
24 Ful beans
25 Black beans
26 Kidney beans
27 Pigeon or Gunga peas
28 Haricot beans
29 Yellow lentils
30 Lima beans
31 Black-eyed beans

raisins and, using the remaining onions, celery and tomatoes, repeat the layers in this order. Mix the vinegar with the liquid from the tomatoes and pour over the casserole. Mix the breadcrumbs with the cheese and some pepper and salt; spread this on top and dot with flakes of butter. Cover and bake in the oven at 375°F, mark 5 for 20 minutes; then lower the temperature to 300°F, mark 2 and continue cooking slowly for about $2\frac{1}{2}$ hours, adding a little water as necessary. Remove the lid 30 minutes before serving to crisp the top. Serve with rough country bread on the cold, wet days my grandmother was fretting about.

6 servings.

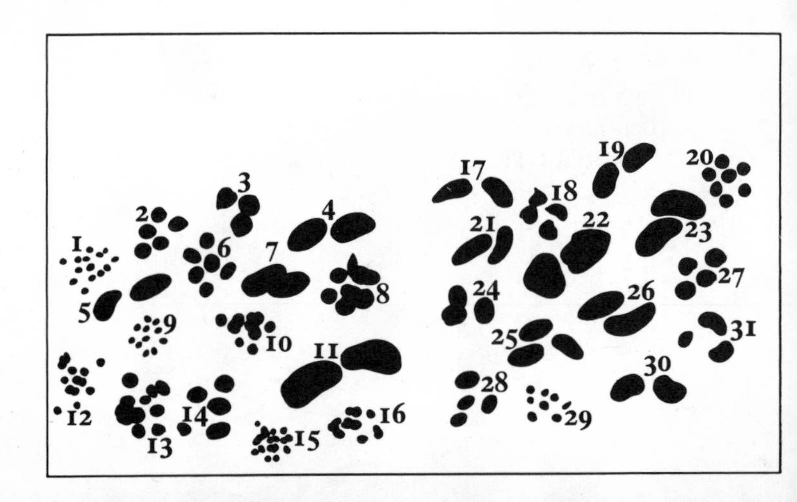

AUSTRIA

8 oz (1¼ cups) white beans
4 oz (½ cup) pearl barley
4 oz pork belly
3 pints (7½ cups) water
1 large onion, chopped
4 sprigs of parsley, chopped
3 oz (6 tablespoons) butter
1 clove garlic, peeled and crushed
salt to taste

Soak the beans and pearl barley together overnight in water. Boil the pork in the fresh water for 40 minutes. Remove the pork and chop into cubes, retain the water.

In a saucepan fry the onion and parsley in the butter until the onion is soft and transparent, then add the strained beans and pearl barley and the garlic, and cook for 3 minutes. Pour in the water and simmer, covered, for about 2 hours. Add the pork and heat thoroughly, then add salt to taste. Serve with dark pumpernickel bread.

4 servings.

Captions to preceeding colour illustrations

Lima stuffed red peppers
Beans (see key on page 56)
Haitian rice & beans

BEEF WITH AUBERGINES & CHICK PEAS

EGYPT

When I do not feel like cooking, I suggest that we eat this stew because my husband knows he cooks it better than I do.

2 medium-sized aubergines (egg plants)
salt
1 large onion, chopped
1 clove garlic, peeled and crushed
6 tablespoons olive oil
1 lb stewing beef, cubed
4 oz (½ cup) chick peas, (soaked, see page 13)
1 large can tomatoes
½ teaspoon coriander powder
salt and pepper to taste
1½ pints (3¾ cups) stock or water

Slice the aubergines, sprinkle with salt and leave them to drain for 15 minutes. Sauté the onion and garlic in 4 tablespoons of the oil in a heavy casserole, until soft. Add the beef and brown quickly on all sides. Stir in the chick peas, salt, pepper and coriander. Cook for 1 minute, pour in the tomatoes with their liquid and simmer for 5 minutes. Stir in the stock or water, cover and cook in the oven at 325°F, mark 3 for 2½ hours. Fry the aubergines lightly in the rest of the oil and add to the stew 15 minutes before the end of cooking time. Serve with warm bread.

4 servings.

JUG-JUG

BARBADOS

I first ate jug-jug in a tiny wooden house in the Scotland district of Barbados. It wasn't until later that I learnt that it is supposed to be a descendant of haggis which came over with some exiled Scots in the 17th century.

½ lb lean pork
¼ lb salt beef
8 oz (1 cup) pigeon peas (soaked, see page 16)
1 large onion, sliced
4-5 stalks celery, thickly sliced
a small handful of parsley, roughly chopped
2 teaspoons oregano
¼ teaspoon caraway seeds
4 tablespoons ground millet
salt and pepper to taste
4 oz (½ cup) butter
chopped parsley to garnish

Cook the meat with the pigeon peas in plenty of water for 1½ hours. Drain, retaining the liquid, and chop the meat and beans together. Put the liquid, onions, celery, parsley, oregano, caraway seeds, millet, salt and pepper into a large saucepan and stir over a low heat for 20 minutes. Add the chopped meat and beans and continue cooking, stirring, for about another 20 minutes until the mixture stiffens. Stir in half the butter and flatten the mixture into a greased oval, ovenproof dish. Bake in the oven at 350°F, mark 4 for 30 minutes, turn out on to a warm serving dish and garnish with chopped parsley and the rest of the butter.

Served on its own with bread, or with barbecued chicken. It makes a good meal for outdoor eating because it has vegetables, grain and meat in one.

4-6 servings.

LAMB WITH BEAN SPROUTS

CHINA

To the Chinese meat usually means pork but they do eat lamb, particularly in the southern areas around Canton. The shoulder and leg are most popular and are cut very fine so that they can be quick fried and stay tender and also because the Chinese consider knives to be barbarous implements and never, never use them at the table.

1 lb lean lamb
1 clove garlic, crushed
1 teaspoon salt
4 tablespoons soy sauce
a pinch of ground ginger
1 teaspoon brown sugar
4 tablespoons oil (peanut or sunflower)
4 spring onions, chopped
4 oz (1 cup) bean sprouts, drained
1 tablespoon cornflour (cornstarch)
$\frac{1}{4}$ pint ($\frac{2}{3}$ cup) water

Cut the meat into thin slivers. Add the crushed garlic to the meat with the salt, soy sauce, ginger, sugar and oil. Mix well and cook gently for 3 minutes. Add the onions and bean sprouts to the pan, mix well and cook for 1 minute (they must be crunchy). Mix the cornflour (cornstarch) to a smooth paste with the cold water, add to the pan and bring to the boil, stirring until slightly thickened.

Serve immediately in a low open dish with steaming bowls of boiled rice for each person. The Chinese sit down to several dishes at the same time but this is difficult to attempt due to the last minute cooking.

It is well to do all the slicing and chopping in advance because it is the preparation of Chinese food that is time-consuming. It is easier to use the Chinese *wok* (a round pan with a curved bottom) but if not, see that the pan used is extremely hot.

4 servings.

See colour picture facing page 104

GOULASH SOUP

HUNGARY

The first goulash soup was made by the horse herdsmen on the Great **Hungarian** Plain in the early 19th Century. They cooked it over an open fire in metal pots and made it with mutton because they could herd sheep along with the horses. Today a typical goulash soup is made with loin of beef.

1 medium-sized onion, chopped
1 oz (2 tablespoons) lard
1 teaspoon paprika
1 lb loin of beef, cubed
1 green pepper, seeded and sliced
2 tomatoes, chopped
3 pints (7½ cups) stock or water
8 oz (1 cup) white haricot beans (soaked see page 14)
4 oz (1 cup) vermicelli or small noodles, optional

Sauté the onion in the lard. Add the paprika and beef and fry until lightly browned. Add the green pepper and tomatoes. Finally, add the stock and beans and simmer for 2 hours. If using vermicelli, add it just before the end of the cooking time so that it remains fairly firm. Serve in deep soup bowls with dark brown or black bread.
6 servings.

CRAB WITH BLACK BEANS

CHINA

Even before modern transport, fresh fish was available all over the vast Chinese continent. Fish from the sea was brought inland alive in water and every village had its well-stocked pond. Freshwater crabs were one of the foods of the Emperors but along the coast the salt-water variety are prized as well. Either can be used with these subtle tasting black beans.

3 oz ($\frac{1}{2}$ cup) black beans (soaked, see page 12)
3 crabs
2 cloves garlic, peeled and chopped
1 tablespoon oil (peanut or sunflower)
1 tablespoon sherry
a pinch of ground ginger

Boil the beans in plenty of water for 1 hour. Remove the meat from the crabs and arrange in a wide, shallow ovenproof dish that will go straight to the table. Put the drained beans, garlic, oil, sherry and ginger in a blender and operate on low speed until they become a smooth paste. Spread this paste over the crabmeat, leaving about 1 inch around the edge uncovered. Cover the dish with a lid or aluminium foil and stand in a shallow baking tin containing water and cook in the oven at 350°F, mark 4 for 45 minutes.

This could be served with the lamb with bean sprouts (see page *60*), on its own as a special delicacy, or with a plate of fried Chinese noodles. If eating with chopstocks, serve in small, individual bowls.

4 servings.

PIGEON PEAS PELAU

TRINIDAD

Since Columbus discovered Trinidad in 1498 there has been a steady stream of people coming to this Caribbean island from all corners of the world, but particularly from the East. All these different sources influence the traditional cooking in this area. Pigeon Peas Pelau is one such traditional dish and is the mainstay of many a family meal.

2 lb chicken, jointed
½ lb lean pork, cut into 2-inch pieces
salt and freshly ground black pepper
1 clove garlic, peeled and finely chopped
1 medium-sized onion, chopped
1 tomato, sliced
6 or 7 blades of chives, chopped
3 tablespoons cooking oil
2 teaspoons sugar
1 lb (2 cups) cooked pigeon peas *(see page 16)*
2 pints (5 cups) water
1 lb (2 cups) rice

Season the chicken joints and pieces of pork with salt and black pepper. Add the garlic, onion, tomato and chives and set aside for about 1 hour. Heat the oil in a saucepan, add the sugar and cook gently until it is brown and bubbly. Add the chicken and pork and cook for 15 minutes, stirring to prevent burning. Add the pigeon peas and water. Bring to the boil, then add the rice. Bring back to the boil, reduce heat and simmer until the rice is tender – about 20 minutes. When almost done, stir once or twice with a fork.

4 servings.

ITALY

This is a spicy, mountain casserole from the Dolomites.

1 lb (2 cups) chick peas (soaked, see page 13)
2 tablespoons olive oil
1 clove garlic, peeled and crushed
1 teaspoon chopped fresh thyme (or $\frac{1}{3}$ teaspoon dried)
a pinch of red pepper flakes
1 lb (4 cups) roughly chopped fennel (green leafy tops and stems)
a piece of salt pork about 1$\frac{1}{2}$ inches square
1 medium-sized onion, sliced

For the tomato sauce
1 medium-sized onion, sliced
2 tablespoons olive oil
1 medium-sized carrot, grated
1 clove garlic, peeled and crushed
1 tablespoon chopped parsley
1 medium-sized can tomatoes
1 teaspoon chopped fresh basil (or $\frac{1}{3}$ teaspoon dried)
salt and pepper
$\frac{1}{2}$ pint (1 cup) meat stock

Put the chick peas on to cook in the water they were soaked in for at least 2 hours, until just tender.

While they are cooking make the tomato sauce. Sauté the onion in the oil until soft and transparent, add the carrot, garlic and parsley and cook for 3 minutes. Add the tomatoes, basil, salt, pepper and meat stock and simmer, covered, for at least 1 hour.

Once the chick peas are cooked, drain them and retain the liquid. In a frying pan heat the oil and add the garlic, thyme, red pepper flakes and fennel. Stir and cook this mixture slowly for 5 minutes. Put the chick peas into a heavy casserole, add the fennel mixture, onion and the tomato sauce with enough of the cooking water from the chick peas to almost cover the peas. Mix well. Place the salt pork in the centre of the casserole, cover and cook in the oven at 300°F, mark 2 for at least 1 hour. Take off the cover for the last 30 minutes of cooking, and add salt to taste just before serving. The completed dish should be quite moist.

Serve with plenty of freshly grated Parmesan cheese and chunky country bread.

6-8 servings.

PEASE PUDDING WITH BOILED BACON

BRITAIN

a large piece of bacon, soaked overnight
1 lb (2 cups) split peas (soaked, see page 17)
1 onion
4 peppercorns
1 oz (2 tablespoons) butter
salt and pepper to taste

Put the bacon in a saucepan of cold water,

bring to the boil and throw the water away.

Tie the drained peas, onion and peppercorns in a muslin cloth. Refill the saucepan with the water, put the muslin bag in with the bacon and bring to the boil again, keeping it on the boil for 30 minutes for each pound of bacon. Take the muslin bag from the water and drain. Put the peas into a bowl (removing the onion and peppercorns) and mix in the butter, salt and pepper. Keep hot.

Remove the rind from the bacon and serve in thick slices with a generous helping of pease pudding.

6 servings.

FABADA ASTURIANA

SPAIN

Asturia is one of four Spanish regions from which much of South America cooking stems. However the tastes are quite distinctive due to the use of saffron and the typically Spanish sausage, chorizo.

12 oz (2 cups) white beans (soaked, see page 17)
½ lb chorizos, sliced
½ lb smoked ham, chopped
½ lb pork, chopped
½ teaspoon saffron
salt and pepper to taste

Put the beans in the bottom of a casserole

or earthenware pot and add the rest of the ingredients. Cover with water by 1 inch, cover the pot tightly and cook in the oven at 350°F, mark 4 for 2½ hours or until the beans are plump and tender. Serve with a tomato and black olive salad.

4-6 servings.

DUCK AND SOYA BEANS

CHINA

This is a dish from the countryside around Peking.

1 medium-sized duck
1½ tablespoon soy sauce
1½ teaspoons cornflour (cornstarch)
4 tablespoons vegetable oil
2 cloves garlic, peeled and sliced
2 shallots, sliced
4 oz (⅔ cup) cooked soya beans *(see page 16)*
¼ teaspoon chilli powder
2 teaspoons sugar
salt to taste

Remove the skin and bones from the duck and cut it into small pieces. Mix together the soy sauce and cornflour (cornstarch) and marinate the duck in this mixture for 10 minutes. Heat the oil in a wok or frying pan and stir-fry the duck with the garlic and shallots for 2 minutes. Sprinkle on a little water (about 2 tablespoons), then add the beans and chilli. Add the sugar and salt and simmer until the duck is tender, about 10 minutes. Serve on its own or with Pork and Bean Sprouts (see page *53*) and boiled rice.

4 servings.

BEAN SPROUT & SOLE SUKIYAKI

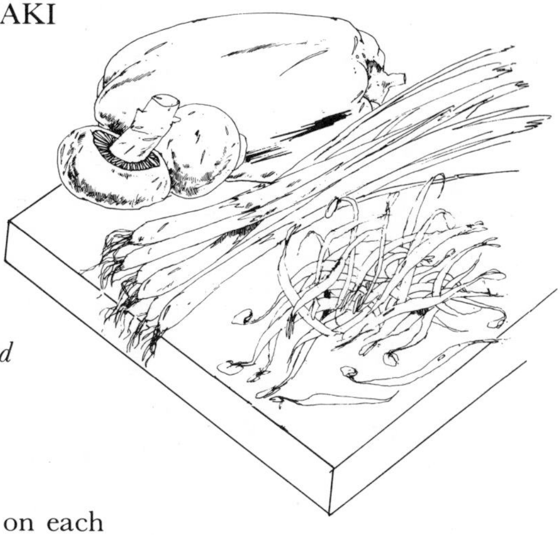

JAPAN

2 fillets of sole
3 tablespoons vegetable oil
2 tablespoons soy sauce
$\frac{1}{4}$ pint ($\frac{2}{3}$ cup) water
$\frac{1}{2}$ green pepper, thinly sliced
4 oz (2 cups) bean sprouts
4 oz ($1\frac{1}{2}$ cups) Chinese cabbage, shredded
6 spring onions, chopped
4 oz (1 cup) mushrooms, sliced
1 tablespoon sugar

Fry the fillets of sole for 1 minute on each side in very hot oil. Pour the soy sauce over and add the water. Bring to the boil. Add the green pepper, bean sprouts, Chinese cabbage, spring onions, mushrooms and sugar and cook for 3 minutes. Serve with boiled rice or boiled egg noodles.

4 servings.

FOOD FOR THE SAINTS

CHINA

A feast for Buddhist saints.

4 tablespoons soya or vegetable oil
2 oz ($\frac{1}{2}$ cup) spring onions, thinly sliced
4 oz (1 cup) mushrooms, thinly sliced
4 oz (1 cup) celery, thinly sliced
2 oz ($\frac{1}{2}$ cup) water chestnuts, thinly sliced
4 oz (1 cup) carrots, thinly sliced
4 oz (1 cup) courgettes, thinly sliced
8 oz (4 cups) bean sprouts
4 oz ($\frac{3}{4}$ cup) Chinese peas (if dried they should be precooked – if unavailable, use garden peas)
8 oz (3 cups) Chinese cabbage, shredded (or ordinary cabbage will do)
4 tablespoons water
1 tablespoon arrowroot
2 tablespoons soy sauce
$\frac{1}{2}$ teaspoon sugar
8 oz bean curd, sliced and browned lightly

Heat the oil, preferably in a wok, otherwise in a large frying pan, add the onions, mushrooms, celery, water chestnuts, carrots, courgettes and stir-fry for 1 minute. Add the bean sprouts, peas and cabbage and stir-fry for a further 1-2 minutes, making sure that the vegetables remain crisp. Add the water, arrowroot, soy sauce, sugar, bean curd and salt. Simmer, covered, for 5 minutes. Remove the lid, turn the heat up again and stir-fry for 1 minute. Serve immediately, accompanied by individual bowls of boiled rice.

6 servings.

FRANCE

A standard dish for those ebullient Sunday restaurant lunches that the French family revel in, from baby to grandmother, to the benefit of the generation gap.

White haricot beans are a good substitute if flageolets are hard to come by.

1 lb (2 cups) green flageolets or white haricot beans (soaked, see page 14)
1 onion, chopped
1 bayleaf
1 leg of lamb
1 clove garlic, peeled and sliced into slivers
2 tablespoons dripping
2 sprigs of fresh rosemary (or ½ teaspoon dried)
salt and pepper to taste
½ cup white wine
butter

Allow at least 1 hour to simmer the soaked beans with the onion and bayleaf. The beans should still be firm, but tender. Drain. Make gashes in the leg of lamb with a sharp knife and insert slivers of garlic under the skin. Spread the dripping over the joint, put the rosemary on top, and season well with salt and pepper. Roast the leg of lamb in the oven at 350°F, mark 4 for 20-25 minutes per pound (if using a meat thermometer, 140°F for rare, 160°F for well done). Baste frequently during cooking so that it is crispy on the outside and pink inside, then put the joint into an ovenproof serving dish. Add the wine to the juice left in the pan and let it boil for 1 minute. Pour this over the drained beans and arrange them around the lamb. Return to the oven for a few minutes to heat thoroughly and add a knob of butter to the beans before serving.

6 servings.

SUCCOTASH

AMERICA

Indian tribes all over America were eating a succotash of corn kernels and lima beans before Columbus arrived. They enriched it with molasses, bear fat and threw in game or fish when they were lucky. Early settlers who were guests of the Indians never quite knew what to expect from the pot.

8 oz (1 cup) lima beans, soaked and cooked for 2 hours (see page 15); or 1 can, drained; or 1 packet frozen lima beans, cooked according to the instructions on the packet
8 oz (1 cup) cooked sweetcorn
2 tablespoons oil
1 teaspoon molasses (honey will do)
salt and pepper to taste
½ teaspoon paprika

Heat the lima beans and corn in the oil in saucepan and add the molasses. Stir until the molasses has melted and the beans and corn thoroughly heated. Season with salt, pepper and paprike. Serve with barbecued chicken and a tomato salad. 4-6 servings.

MOORS & CHRISTIANS

CUBA

Moors and Christians – black beans and white rice – is the food Cuban tobacco and sugar plantation workers have come home to for generations. Cooked under the stars on a night that is going nowhere except deeper into the shadows, the glistening black beans in the pot seem to encapsulate some of the magic of the Caribbean.

8 oz (1 cup) black beans (soaked, see page 12)
8 oz (1 cup) rice
1 medium-sized onion, chopped
1 large clove garlic, peeled and crushed
1 small green pepper, chopped
salt and pepper to taste
a bunch of fresh parsley, chopped
1 tomato, chopped
½ onion, finely chopped

Put the beans, chopped onion, garlic and green pepper in an iron pan with plenty of water. Cover and simmer for 2 hours. Drain and serve with the parsley, tomato and onion chopped on top, with fluffy white rice in another dish.

4 servings.

COMMUNE BEANS & BEAN CURD

CHINA

This is a dish from Central China which doesn't find its way on to the tables of the specialized Cantonese, Pekinese and Shanghai restaurants that we know in the west. By Chinese standards it is simple fare, its main ingredient being the bean curd cakes made from the soya bean and purchased from any Chinese food shop.

½ lb pork
6 bean curd cakes
4 tablespoons vegetable oil
1 medium-sized onion, chopped
⅓ of a Chinese cabbage, shredded
3 teaspoons salted black beans
⅛ teaspoon dried chillies
4 tablespoons soy sauce, diluted with the same amount of water
1 teaspoon sesame seeds
2 teaspoons sesame or peanut oil

Slice the pork as thinly as possible into pieces about 1 inch square. Cut the bean curd slightly smaller, about ¾ inch square. Fry the bean curd on all sides in 2 tablespoons of hot oil in a wok or frying pan, and then remove from the pan. Add the rest of the oil and fry the pork quickly on both sides. Stir-fry the onions for 1 minute followed by the cabbage, black beans and chillies for 2 minutes. Pour the diluted soy sauce into the wok and stir-fry for 2 minutes. Carefully fold in the bean curd.

Fry the sesame seeds in very hot sesame

or peanut oil. Serve the beans and bean curd in a deep bowl with the fried sesame seeds sprinkled on top and accompanied by a bowl of plain boiled rice.

6 servings.

FUL MEDAMES *NILE BEANS*

EGYPT

For centuries this has been the food of the Egyptian peasant. It found its way from the mud-brick houses on the banks of the Nile to the salons of Cairo, without changing its ingredients. Ful are warm, comforting and very filling.

1½ lb (3 cups) ful beans (soaked, see page 14)
1½ tablespoons ground cumin
pepper
4-6 cloves garlic, peeled and crushed
salt
4 hard-boiled eggs (warm)
4 onions, roughly chopped
2 lemons, quartered
olive oil, preferably unrefined heavy green olive oil
salt and pepper to taste

Cover the soaked beans with water by 2 inches, add the cumin, pepper and garlic and cook in the oven at 300°F, mark 2 for about 3 hours until the beans are very soft and the remaining juice has thickened. Salt to taste 15 minutes before the end of cooking time.

Bring a bowl brimming full of these beans to the table and set it amongst the bowls of warm hard-boiled eggs, the raw onions, quartered lemons and a jug of olive oil. Then the ritual begins. Invite your guests to crumble their own eggs on top of the beans, add onions, squeeze on the lemon juice and pour on enough oil to give the beans a smooth consistency.

Serve with hummus (see page *40*) and salati baladi (peasant salad) (see page *153*).

The whole meal is eaten traditionally by scooping it up with warm Arab bread (see page *157*)– in my family, we use a fork and bread together. Ful are just as delicious cold, eaten the next day as a salad mixed with a generous amount of chopped parsley and coated in a yoghurt, oil and vinegar dressing. A tomato and onion salad is a perfect accompaniment.

4 servings.

See colour picture facing page 105

CHILLI CON CARNE

MEXICO

When I came to live in England from California at the age of eight, I couldn't understand why my new schoolfriends didn't like Chilli Con Carne for tea. Today, this Mexican classic of the bean world is the one dish everyone mentioned when I said I was going to write a bean book. The varieties are endless – the meat can be cubed or minced, the beans mixed with the sauce or separate, sophisticates cook their meat in a bottle of red wine and Peruvians serve the lot on rice. This recipe is the one my mother found in the early forties in Mexico. The only difference perhaps being that she serves it hotter and hotter every year!

2 lb (3 cups) kidney beans
2 oz (4 tablespoons) lard
¼ cup olive oil
2 lb lean beef (cut in ¾-inch cubes)
2 large onions, finely chopped
3 cloves garlic, peeled and crushed
1 tablespoon paprika
2-3 tablespoons chilli powder
2 teaspoons oregano
salt and pepper to taste
1 large can tomatoes
½ pint (1¼ cups) water
1 large onion, coarsely chopped, to garnish

Cook the kidney beans in plenty of water for 2 hours and drain. Heat the olive oil and lard, add the meat and cook until brown all over. Add the two finely chopped onions and garlic and cook until transparent. Mix in the paprika, chilli powder, oregano, salt and pepper, stir for a minute or two and then add the tomatoes and water. Simmer until the meat is tender, about 1 hour, add the kidney beans and simmer at least 15 minutes more. Serve in bowls with the coarsely chopped onion on top; melba toast and a tomato salad with a light lemon dressing at the side.

6 servings.

BOSTON BAKED BEANS

AMERICA

The Indians, in what later became New England, pit-cooked their beans in bean holes, in maple sugar with a lump of bear fat. Settlers replaced the maple sugar with molasses, the bear fat with salt pork and retained the principle of long slow cooking.

¼ lb lean salt pork
1 small onion, sliced
1 lb (1½ cups) red beans (soaked)
2 tablespoons sugar
1½ tablespoons molasses
½ teaspoon dry mustard
salt to taste

Soak the salt pork in boiling water for 15 minutes, drain, cut off the rind and cut into small pieces. Put the onion and pork into an earthenware pot. Add the drained beans, sugar, molasses and mustard. Pour in water to 1 inch above the beans and bake covered in a low oven, 325°F, mark 3, for about 6 hours, taking care to keep enough liquid in the pot. Season to taste 15 minutes before serving. Serve with whole-wheat bread with cornmeal (see page *156*).

6 servings.

PASTA E FAGIOLI *PASTA & BEANS*

ITALY

Often called 'thunder and lightning' this dish came about as a way of using the small broken pieces of noodles left in the bottom of the sacks and barrels in shops.

6 slices bacon, cut into 1-inch strips
1 tablespoon olive oil
1 medium-sized onion, chopped
1-2 cloves garlic, peeled and crushed
2 stalks celery, chopped
1 tablespoon tomato purée
1 large can tomatoes, chopped
1 lb (2½ cups) cooked chick peas (see page 13)
3 tablespoons chopped parsley
1 teaspoon chopped fresh basil
salt to taste
½ pint (1¼ cups) water
8 oz (2 cups) noodles, in small pieces

Cook the bacon slowly in the oil until brown. Remove and drain. Sauté the onion, garlic and celery lightly, then simmer with the tomatoes and their liquid for 5 minutes. Return the bacon to the pan and add the chick peas, parsley, basil, salt and water. Cover and cook over low heat for 30 minutes.

Boil the noodles separately until tender and stir into the chick pea mixture. Serve with a bowl of freshly grated Parmesan cheese.

A simpler version can be made by just mixing together cooked chick peas, noodles and Parmesan topped with oil or butter. The simpler version is a good standby meal for summer or winter because the ingre-cients all come from the store cupboard.

4 servings.

MUNG BEAN CURRY

SRI LANKA

Most peas and beans on the island are cooked into curries which, along with rice, are the staple food of the Sri Lankans. The spices for which the island has been internationally famous for 2000 years are the dominant ingredients and can be cut by as much as half if a mild version is preferred.

8 oz (1 cup) mung beans (soaked overnight)
2 tablespoons oil
1 large onion, chopped
1 green chilli, chopped
3 teaspoons chilli powder
1 teaspoon coriander powder
1 teaspoon cumin powder
¼ teaspoon tumeric powder
1 pint (2½ cups) coconut milk
salt to taste

Roast the mung beans slightly in a hot oven and spin in the blender to chop them up. Heat the oil in a pan and add the onions and cook until light brown. Add the mung beans with all the other ingredients except the coconut milk and keep frying for 5-7 minutes. Add the coconut milk, bring to the boil and simmer for 40 minutes. Serve in a low, open bowl with a large amount of boiled rice.

8 servings.

CASSOULET

FRANCE

The Cassoulet, as the French call it to show their respect, comes from the farmhouse kitchens of the Languedoc, south-west France, where abundant quantities of meats and fowl are to be found hanging in the larder. Traditionally and practically, it should be cooked for at least 8 very hungry people with cold winter appetites.

2 lb (4 cups) white haricot beans (soaked, see page 14)
3 medium sized onions, chopped
8 oz (12 slices) bacon, chopped
4 cloves garlic, peeled and crushed
2 tomatoes, chopped
½ teaspoon chopped thyme
4 teaspoons chopped sage
1 teaspoon chopped parsley
2 pints((5 cups) meat stock
½ fresh goose (or a whole duck or a chicken)
1 lb coarse pork sausages
½ lb breast of lamb
8 oz (2 cups) fresh breadcrumbs

Cook the beans for 1 hour in the water they soaked in. Fry the onions with the bacon, add the garlic, tomatoes and herbs and simmer with the stock for 15 minutes. Cut all the meat into chunky pieces and put on the bottom of a large earthenware pot together with the onions and bacon mixture. Put the strained beans on top, pour in the stock, bring to boil on top of the oven and spread a thick layer of breadcrumbs on top. Allow to cook slowly in the oven at 350°F, mark 4 for about 1 hour, uncovered, until the breadcrumbs are crispy and the beans are tender. Serve in its own pot with a light, green salad and goat cheese to follow.

8 servings.

SPICED CHICKEN

ANTIGUA

Called groundnuts in the Caribbean, peanuts are often pounded into powder and added to sauces and casseroles which they thicken to make delightfully rich.

2 oz (4 tablespoons) butter
3 lb chicken, cut in pieces
1 clove garlic, peeled and chopped
1 tablespoon chopped chives
½ teaspoon thyme
1 bayleaf
salt and pepper to taste
½ teaspoon nutmeg
½ teaspoon cinnamon
6 oz (1½ cups) ground peanuts
¾ pint (2 cups) water

Heat the butter in an iron or similar heavy casserole. Quickly fry the chicken pieces on all sides. Add the garlic and sauté until golden; then add the chives, thyme, bay-leaf, salt, pepper, nutmeg, cinnamon and ground peanuts. Pour in the water, cover and simmer for 1 hour.

Serve with boiled sweet potatoes or rice, and a chopped celery and raw cabbage, salad, garnished with sliced green peppers. 6 servings.

CHICKEN LIVERS & SNOW PEAS

CHINA

Snow peas are a delicacy to the Chinese and are stir-fried quickly to keep their delicate flavour and crispness. Our equivalent in the West are mange tout. Snow peas are available from time to time in Chinese stores and our own supermarkets. Because the dish is quick to prepare and cook, I find it very practical to have it at the end of the working day.

½ lb snow peas, topped and tailed
2 tablespoons dry sherry
2 tablespoons soy sauce
½ teaspoon sugar
½ teaspoon cornflour (cornstarch)
½ lb chicken livers, thinly sliced
3 tablespoons vegetable oil
¼ inch of ginger root, chopped
salt to taste

Blanch the snow peas in boiling water for 1 minute to start them cooking and bring out their greenest colour. Run cold water over them in a sieve.

Mix the sherry, soy sauce, sugar and cornflour in a bowl with a fork, add the chicken livers and let them soak in the mixture.

Heat half the oil in a wok or frying pan over a high heat, add the peas and stir-fry for 2 minutes. Remove, add the rest of the oil and stir-fry the ginger for 1 minute. Lower the heat and stir-fry the chicken livers for 2 minutes. Put the peas back into the pan, season with salt and stir-fry for 30 seconds. Serve with boiled rice and Fried Bean Curd (see page *47*).

2 servings.

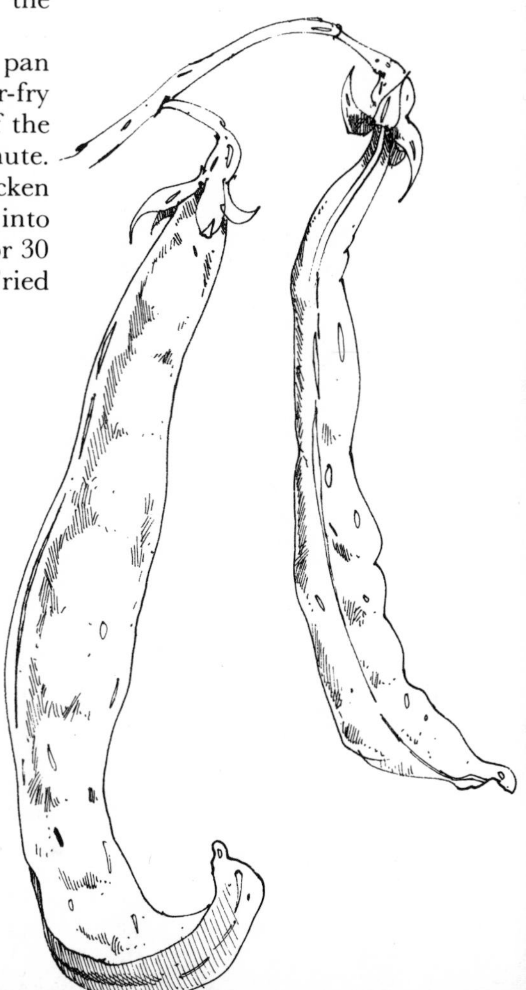

DIZA *LAMB AND BEAN STEW*

IRAN

Going to see the carpet washers in Iran is a shocking experience in irreverence. I arrived at lunchtime, only to be invited to share this communal bean purée, stewed mutton and broth spread out on a carpet that looked as though it belonged in the rather more rarified atmosphere of a Shah's Palace. The carpet probably has gone to some grand home or other but at that time it was doing a great job as a picnic blanket/tablecloth.

2 lb stewing lamb, cubed
about 1 lb lamb bones
1 large onion, thickly sliced
6 oz (1 cup) dried broad beans (soaked, see page 12)
6 oz (1 cup) chick peas (soaked, see page 13)
1½ pints (3¾ cups) water
1 small can tomatoes, chopped
juice of 1 lemon
½ teaspoon coriander
½ teaspoon tumeric
salt and pepper to taste
1 medium-sized onion, finely chopped

Put the lamb and lamb bones on the bottom of an iron casserole, spread the onions on top, then the broad beans and chick peas. Pour in the water and bring to the boil, skimming the surface to remove the scum. Reduce the heat, cover and simmer for 1 hour.

Add the tomatoes, lemon juice, coriander, tumeric, salt and pepper and simmer, covered, for a further $1\frac{1}{2}$-2 hours or until the chick peas and broad beans are tender. The ingredients should be covered all the time during cooking and more water may have to be added.

Sieve out the lamb and bones (discarding the bones), strain the broth into a soup tureen and mash the beans to a rough purée. Mix the raw chopped onion into the purée, spoon it on to a big shallow dish and lay the lamb pieces on top. Serve the soup first in small individual bowls, followed by the beans and meat.

A thirst-quenching Persian drink usually drunk with Diza is *dugh* – made by diluting yoghurt with water and adding a little salt.

4 servings.

GUNGA PEAS & RICE

BAHAMAS

You won't find this dish on the tables of the superb, luxury hotels of the Bahamas but on every other table on the 30 inhabited islands of this 700 island sun-blessed archipelago.

¼ lb bacon, chopped in small cubes
6 tablespoons (½ cup) cooking oil
1 lb (4 cups) chopped onions
½ green pepper, chopped
5 oz (½ cup) tomato paste
1 medium-sized can tomatoes, drained and chopped
2 teaspoons chopped fresh thyme (or ½ teaspoon dried)
black pepper to taste
1 lb gunga peas (soaked, see page 16)
2 lb (4 cups) rice
salt to taste

Fry the bacon in the oil in an iron pot until cooked. Add the onions and green pepper and fry until soft. Add the tomato paste, tomatoes, thyme and black pepper and cook for 5 minutes. Stir in the peas. Add the juice drained from the can of tomatoes to the water the peas were soaked in and make the quantity up to 3½ pints (9 cups) and pour over the beans. Bring to the boil. Sprinkle the rice in, stirring all the time and reduce the heat to very low. Cover and cook for about 30 minutes until the rice is done. Serve as an accompaniment to barbecued chicken.

8 servings.

CHICK PEAS & PAPAYA CURRY

INDIA

Into this exquisite curry go the crunchy small Indian chick peas which provide such a marvellous contrast to the soft, juicy, fragrant papayas.

2-3 papayas (papaws) – or 1 medium-sized can, cubed
4 oz (⅔ cup) cooked chick peas *(see page 13)*
6 curry leaves
3 tablespoons desiccated coconut
2 green chillies, chopped
½ teaspoon cumin
¼ teaspoon turmeric
salt to taste
4 small onions, sliced
½ oz (1 tablespoon) butter

Put the papaya pieces and chick peas into a saucepan with 3 of the curry leaves and bring to the boil. Skim the surface of the water and then drain the papaya and chick peas.

Mix the coconut with the chillies, cumin, turmeric, salt and enough water to make a gravy, and simmer until thick. Add the other 3 curry leaves and the onions and butter, and simmer for 5 minutes. Pour the sauce over the papayas and chick peas and serve with plain boiled rice.

2 servings.

HONG KONG OMELETTE

HONG KONG

What better place for East to meet West than in the frying pan? And it is a quick and nutritious occasion.

3 eggs
1 tablespoon milk
salt and pepper to taste
3 spring onion tops, chopped
6 oz soya bean sprouts
1 tablespoon soya or vegetable oil

Beat the eggs. Mix in the milk and salt and pepper, followed by the onion tops and bean sprouts. Heat the oil in an omelette or medium-sized frying pan and cook quickly.

2 servings.

NEW ZEALAND

This is one of my grandfather's recipes.

2 lb (5 cups) young broad beans
1 large onion, finely chopped
½ teaspoon dried sage
¼ teaspoon Marmite or yeast extract
3 tablespoons wholehweat flour
2 tablespoons milk
2 egg yolks, well beaten
2 tablespoons grated cheese
2 oz (1 cup) fresh wholewheat breadcrumbs

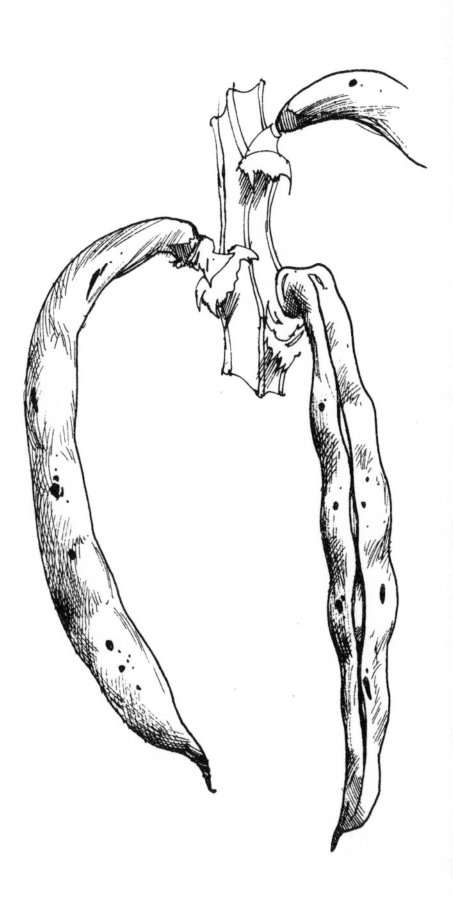

Cook the beans, onion and sage together in a saucepan, allowing just enough water to make a little gravy when done. When the beans are nearly cooked, remove from the heat and add the Marmite or yeast extract to the juice and turn into a greased pie dish.

Mix the flour with the milk until of a smooth consistency, add the egg yolks and cheese and stir well. Spread this creamy mixture over the beans in the pie dish, add a layer of breadcrumbs and bake in the oven at 350°F, mark 4 for 30 minutes.

4 servings.

LIMA STUFFED RED PEPPERS

6 red peppers of even size
1 oz (2 tablespoons) butter
1 medium-sized onion, finely chopped
1 clove garlic, peeled and crushed
¼ teaspoon chilli powder
4 oz (1 cup) mushrooms, chopped
1 tablespoon finely chopped parsley
¼ pint (⅔ cup) stock (vegetable or meat)
3 tablespoons tomato purée
salt and pepper to taste
6 oz (1 cup) cooked lima beans, chopped
4 oz (½ cup) ham, finely chopped

Blanche the peppers for 5 minutes in salted water, drain, cut off the tops and scoop out the seeds. In a saucepan melt the butter and soften the onion with the garlic and chilli powder, then add the mushrooms. Sauté this mixture lightly then add the parsley, stock, tomato purée and seasonings and bring to the boil. Reduce the heat, cover and simmer for about 15 minutes. Drain off and retain the surplus liquid. Add the limas and ham to the mixture. Fill the peppers and replace the tops. Brush with melted butter and arrange in an oiled ovenproof dish. Dilute the surplus liquid until it just covers the bottom of the dish. Cook in the oven at 350°F, mark 4 for 30 minutes. Serve trimmed with parsley.

6 servings.

See colour picture facing page 56.

PERSIAN MILLET & LENTILS

IRAN

This centuries-old recipe is a good example of early nutrition without science, of a people combining two of their basic food stuffs which, incomplete in proteins alone, combine to provide an extremely healthy diet.

2 tablespoons oil
1 onion, chopped
1 clove garlic, peeled and crushed
3 pints (7½ cups) water
3 cloves
12 ground cardamom seeds
½ teaspoon cinamon
¼ teaspoon mace
2 bayleaves
8 oz (1 cup) lentils (soaked for 1 hour)
8 oz (1⅓ cups) millet (soaked overnight)

Heat the oil in a large iron casserole dish. Sauté the onion until clear, then add the garlic. Add the water and spices and bring to the boil. Add the lentils and millet. Cover and simmer for about 2 hours. Drain off the remaining liquid and cook, covered, in the oven at 300°F, mark 2 until all the moisture has evaporated. Serve with a cucumber and yoghurt salad (see page *151*).

6 servings.

INDIAN KEDGEREE

This is the original dish from which the traditional British Army breakfast evolved.

1 large onion, sliced
1½ tablespoons olive oil
2 teaspoons curry powder
8 oz (1 cup) rice
8 oz (1 cup) red lentils
¾ pint (2 cups) chicken stock
1 bayleaf
salt and pepper to taste
2 large hard-boiled eggs, sliced
chopped parsley
lemon quarters

Sauté the onion in the olive oil, add the curry powder, stir in the rice and cook for 2-3 minutes. Transfer this to a casserole, add the lentils, stock, bayleaf and seasoning. Cover and cook in the over at 350°F, mark 4 for about 40 minutes until all the stock is absorbed. Serve steaming hot, garnished with the slices of hard-boiled egg and chopped parsley and surrounded with quarters of lemon to squeeze on top.

It should be accompanied with cucumber and yoghurt salad (see page *151*) and chutney. It makes a delicious meal on its own but is also a good accompaniment for a meat or chicken curry.

4 servings.

GERMANY

1 small red cabbage, shredded
3 onions, sliced
1 lb cooking apples, quartered
12 oz (2 cups) cooked kidney beans
¼ teaspoon allspice
¼ teaspoon cinnamon
¼ teaspoon thyme
2 cloves garlic, peeled and crushed
1 teaspoon finely grated orange rind
salt and pepper to taste
2 tablespoons brown sugar
½ pint (1¼ cups) red wine
2 tablespoons wine vinegar

Cook the cabbage for 5 minutes in a covered pan, then drain.

Take a deep casserole and layer the ingredients, beginning with cabbage, then onions, apples, beans; seasoning each layer with the spices, thyme, garlic, orange rind, salt and pepper. Sprinkle brown sugar over the top, then add the wine and wine vinegar, diluted with a little hot water. Cover and simmer very slowly in the oven at 325°F, mark 3 for 45 minutes or until tender, but not too mushy, adding a little more wine if necessary. Serve with pumpernickel bread and cream cheese.

6 servings.

PREYBRANATZ *ONION & BEAN CASSEROLE*

YUGOSLAVIA

Casseroled beans are traditionally the basic diet of the peasants and there are many variations with and without meat. This Serbian recipe is from the wine-growing area around the Danube. You have to love onions to enjoy this one because it calls for half beans/half onions.

5 medium-sized onions, sliced
3 tablespoons oil
1 lb (2 cups) brown borlotti beans (soaked, see page 12)
1 bayleaf
paprika to taste (at least 1 teaspoon)
salt to taste

Sauté the onions in the oil until they are soft and transparent. Put the drained beans, bayleaf, paprika and onions into an earthenware pot. Cover with water by 2 inches. Cover the pot and bake in the oven at 325°F, mark 3 for at least 2 hours. Check periodically that all the juice isn't boiling away and add more water if necessary. Add salt to taste 15 minutes before serving.

Serve with a lettuce and cucumber salad dressed lightly with oil and vinegar and wholemeal bread.

6-8 servings, depending on whether it is served as a first course or a supper dish

HUNGARY

This really is a complete meal in itself. It is at its best served for supper on a cold winter's evening.

½ lb chicken livers, chopped
4 oz (½ cup) potatoes, chopped
1 oz (2 tablespoons) fat
1 oz (¼ cup) flour
4 oz (⅔ cup) cooked white beans
½ teaspoon sugar
2½ pints (6¼ cups) water
a handful of chopped parsley

For the Borzo Teszta
2 oz (½ cup) flour
a pinch of salt
1 egg, beaten
about ½ pint (1 cup) milk or milk and water
fat for frying

Brown the livers and potatoes in the fat. Sprinkle with the flour and cook for 2-3 minutes, then add the beans, sugar and water. Simmer for 30-40 minutes, adding the parsley at the last minute. To make the Borzo Teszta, mix the flour and salt into a bowl, make a well in the centre and pour in the egg and half the liquid. Beat with a wooden spoon until smooth. Gradually add a little more liquid until the mixture is of a slightly thicker consistency than a pancake mixture.

Heat the fat in a frying pan until it is

smoking hot. Drop the mixture into the fat in teaspoonfuls and fry until golden brown. Remove from the pan and drain. Serve these separately in a hot dish and sprinkle them into each bowl of soup.

4 servings.

BROAD BEANS IN EGG CUSTARD

SCANDINAVIA

A favourite all over Denmark, Sweden, Norway and Finland in mid-summer when the broad beans are fresh.

1 lb (2¼ cups) fresh broad beans
3 eggs
1 pint (2½ cups) milk
salt and pepper to taste
1 tablespoon chopped parsley

Boil the broad beans in salted water until tender. Drain and place in a pie dish. Beat the eggs with the milk, season with salt and pepper and parsley and pour over the beans. Cook in the oven at 350°F, mark 4 until the custard is set and browned on top (about 20 minutes).

4 servings.

ITALY

Here is a beautiful dish of speckled red borlotti beans and warm, brown spices that wafts wonderful smells through the house as it is cooking.

a small bunch of parsely, chopped
2-3 cloves garlic, peeled and crushed
½ teaspoon grated nutmeg
½ teaspoon ground cloves
freshly ground black pepper
12 rashers of bacon
1 lb (2 cups) borlotti beans (soaked, see page 12)
salt to taste.

Mix the parsley, garlic, nutmeg, cloves and pepper together and sprinkle on to the bottom of an earthenware pot. Lay the slices of bacon on top. followed by the beans and enough water to cover by 1 inch.

The slower this dish cooks, the more the beans absorb the spices, so put the tightly covered pot into the oven at 325°F, mark 3 at breakfast time and it will be ready for lunch 3 or 4 hours later. Or start them after lunch to be ready for supper. Add salt to taste just before serving in soup plates.

This is a good weekend meal as it takes little time to prepare, can be made economically for any amount of people and can be left in the oven while the family is out.

4-6 servings.

SEKIHAN *RED BEANS WITH RICE*

JAPAN

This is a festive Japanese dish which uses the colour of the water in which the red aduki are cooked to colour the rice. Typically it is made with sweet, soft rice but half a teaspoon of caster sugar added to ordinary rice cooked until fairly soft will do very well.

6 oz (¾ cup) aduki beans (soaked, see page 12)
1 lb (2 cups) rice
½ teaspoon caster sugar
2 teaspoons sesame seeds
salt to taste
2 tablespoons vegetable oil

Keep the liquid the aduki beans were soaked in and use it to soak the rice for 2 hours. Mix the aduki beans and rice, add the sugar and steam for 1 hour either in a bamboo steamer or in a sieve or colander over boiling water. Mix the sesame seeds and salt and fry in the oil. Serve the beans and rice in individual bowls with the sesame seeds and oil in a small dish for flavouring at the table. This dish is particularly good with fish baked in the oven with diluted soy sauce and a little garlic.

4-6 servings.

FRIJOLES REFRITOS

MEXICO

Refried beans are macho food – on the grounds that if you can eat them as the Mexicans do for breakfast, lunch and dinner you must be made of stern stuff. However, delicious they are and once tried for Sunday brunch, it will never seem the same without them.

1 piece salt pork or streaky bacon
2 teaspoons cumin
7 oz (1 cup) pinto or pink beans (soaked, see page 16)
bacon fat or lard
salt to taste
4 oz (1 cup) grated cheese

Chop the salt pork or streaky bacon and sprinkle with cumin. Put it into an iron pot with the drained beans and enough water to cover by 1 inch. Bring to the boil, cover and simmer for about 2 hours until the beans are tender and there is not much liquid left. Heat a little bacon fat or lard in a frying pan and sauté the beans, mashing some with a fork and leaving the rest whole. Turn into a flat ovenproof dish, sprinkle salt and cheese over the top and melt the cheese under the grill.

Serve with fried eggs and grilled tomatoes. Make plenty of Frijoles as the flavour improves over a couple of days.

4 servings.

BELGIAN BEANS

BELGIUM

When I was a student in Spain, I shared a room with a Flemish girl called Magda who didn't like Spanish food at all and wouldn't let me go shopping because she wasn't all that impressed with boiled eggs either.

So we made an arrangement and it turned out to be a wildly successful one. She cooked, I read dramatic Spanish poetry out loud and we became so popular that in the end we had to charge our friends for their meals. This was one of Magda's favourites, or was it because she knew I liked boiled eggs?

1½ lb (3 cups) haricot beans (soaked, see page 14)
3 large tomatoes, sliced
¾ pint (2 cups) white sauce
a handful of chopped fresh parsley
6 oz (3 cups) fresh breadcrumbs
3 hard-boiled eggs, chopped
1 oz (2 tablespoons) butter
salt, pepper and paprika to taste

Cook the haricot beans in plenty of water for 1 hour. Put the tomatoes into a greased casserole dish and pour in enough white sauce to cover. Sprinkle generously with parsley, add a layer of breadcrumbs, then chopped eggs and then the beans. Season with salt and pepper. Pour on the rest of the white sauce, topping with the remaining breadcrumbs and dot with butter. Sprinkle a little paprika on top – about as much as looks attractive will be enough for the flavour. Serve with French bread.

4 servings.

ITALY

A spicy, inexpensive and quick dish, especially if tinned chick peas are used.

8 oz (1 cup) chick peas (soaked, see page 13)
1 medium-sized onion, chopped
1½ tablespoons olive oil
2 cloves garlic, crushed
4 anchovies, roughly chopped
1 large can tomatoes
1 bayleaf
a pinch of saffron
a pinch of cinnamon
salt and pepper to taste
7 oz (1 cup) rice
chopped parsley to garnish

Cook the chick peas for 2 hours or if using tinned, empty and drain. Sauté the onion in the oil in a saucepan until soft; add the garlic, anchovies, tomatoes, bayleaf, saffron and cinnamon. When the mixture becomes a thick purée (10 to 15 minutes) add the drained chick peas, season with salt and pepper and simmer gently for 15 minutes. Serve as a supper dish, putting the boiled rice on the plates first, chick peas and purée on top and sprinkle chopped parsley over both.

4 servings.

PEANUT SOUFFLE

$\frac{3}{4}$ oz ($1\frac{1}{2}$ tablespoons) butter
6 tablespoons flour
2 oz ($\frac{1}{4}$ cup) peanut butter *(see page 45)*
1 oz (2 tablespoons) finely ground peanuts
$1\frac{1}{2}$ teaspoons salt
1 teaspoon lemon juice
$\frac{1}{4}$ pint ($\frac{2}{3}$ cup) scalded milk
4 egg yolks, beaten
4 egg whites, beaten stiff

Melt the butter and quickly add the flour, peanut butter, ground peanuts, salt, and lemon juuice. Stir over a low heat for 3 minutes. Pour the scalded milk in slowly, stirring all the time and raise the heat until the mixture reaches boiling point. Pour the mixture into a bowl containing the beaten egg yolks, mix well and cool. Fold in the egg whites and spoon into a greased soufflé dish. Put the soufflé dish into a pan of water and cook in the oven at 375°F, mark 5 for 30 minutes. Serve immediately it is cooked with a watercress and orange salad dressed with oil and lemon or a celery and apple salad with oil and lemon.

4 servings.

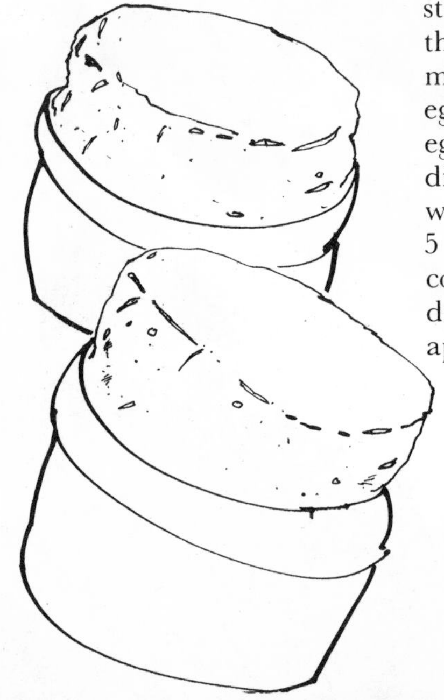

PEAS AND SAUERKRAUT IN THE OVEN

GERMANY

This comes from the south west corner of Germany, along the Rhine.

6 oz (1½ cups) sauerkraut
2 oz (½ cup) mushrooms, sliced
½ pint (1 cup) sour cream
¾ pint (2 cups) water
salt and pepper
8 oz (1 cup) whole dried peas (soaked, see page 15)
paprika

In a saucepan, put the sauerkraut, mushrooms, sour cream, water, salt and pepper and simmer for 1 hour, stirring towards the end of cooking time as the liquid reduces. Grease the base and sides of a casserole and spoon in a layer of half the sauerkraut mixture, followed by the peas, then the remaining sauerkraut mixture. Add the liquid left in the saucepan and sprinkle the top with paprika. Cook, uncovered, in the oven at 350°F, mark 4 for 30 minutes.

4 servings.

Captions to preceeding colour illustrations

Lamb with bean sprouts
Bean puree with yoghourt & Three bean salad
Ful Medames

CALIFORNIA PEANUT LOAF

AMERICA

This is a marvellous supper dish that my mother often used to make for us.

4 oz (2 cups) fresh wholewheat breadcrumbs
1 oz (2 tablespoons) butter
4 oz (1 cup) chopped peanuts
4 oz (1 cup) carrots finely grated
4 oz (1 cup) celery, thinly sliced
2 oz (1½ cups) parsley, finely chopped
1 tablespoon single cream
2 eggs, beaten
salt to taste

Mix 1 oz (½ cup) of the breadcrumbs with half the butter and grease a cake tin with this mixture. Mix the rest of the breadcrumbs with all the other ingredients (except the remaining butter). Turn the mixture into the tin and press it down slightly. Dot the top with the rest of the butter. Place a piece of greaseproof paper on top and bake in the oven at 350°F, mark 4 for 30 minutes. Turn out on to a warm plate.

This is delicious served with a chopped apple and yoghurt salad.

4 servings.

BLACK-EYED BEANS & HOG JOWL

AMERICA

The superstition in Texas runs that if you eat black-eyed beans on new Year's Day you will have a lucky year. This recipe will be found all over the South and its origins are from plantation days when the plantation owners slaughtered their pig's and gave the hog jowl and pigs' tails to their farm-hands.

1 lb (2 cups) black-eyed beans (soaked, see page 12)
8 oz (12 slices) bacon, chopped
freshly ground black pepper
2 medium-sized onions, chopped
salt to taste

Keep enough of the soaking water to cover the black-eyed beans by 1 inch in a heavy pan. Add the bacon and freshly ground pepper. Bring to the boil and simmer, covered, for $1\frac{1}{2}$ hours. Add the onions and salt to taste. Add more pepper if necessary. Simmer until the liquid has reduced to a syrupy consistency and the meat and beans are very soft (about 1 hour). Serve with corn or wholemeal bread and a green salad with a sharp vinegar dressing.

4 servings.

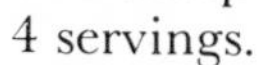

SUPER LOAF

4 oz (2 cups) fresh wholewheat breadcrumbs
1 teaspoon marjoram
1 teaspoon sweet basil
1 teaspoon cumin
½ teaspoon salt
1 egg, lightly beaten
6 oz (1 cup) cooked soya beans, mashed slightly
6 oz (1 cup) cooked brown rice
3 oz (¾ cup) chopped peanuts
4 oz (1 cup) grated mild Cheddar cheese
4 oz (1 cup) finely chopped mushrooms
2 tablespoons finely chopped parsley
1 clove garlic, peeled and crushed
1½ teaspoons sesame seeds
½ oz (1 tablespoon) butter

Soften the breadcrumbs with the dried herbs, cumin and salt. Add the egg and mix well. Add the remaining ingredients (except the sesame seeds and butter) and form into a loaf shape. Place on a flat, oiled, baking dish. Sprinkle with sesame seeds and dot with butter. Cook in the oven at 325°F, mark 3 for 30 minutes, and baste with a little hot water or stock.

6 servings.

CONGRIS

CUBA

This is a quickly prepared, attractive, red bean and white rice dish. A variation retains the liquid which the beans have been cooked in, and uses this to cook the rice, so that the whole dish is a warm red.

1 medium-sized onion, finely chopped
1 clove garlic, peeled and crushed
1 oz (2 tablespoons) lard
2 tomatoes, peeled and chopped (or 1 small can)
1 green pepper, finely chopped
1 lb (2 cups) cooked kidney beans
6 oz (1 cup) long-grain rice
$\frac{3}{4}$ pint (2 cups) water
salt and pepper to taste

Sauté the onion and garlic in the lard in a heavy casserole, until soft and transparent. Add the tomatoes and green pepper and cook, stirring, until the mixture thickens. Add the beans, rice, water and salt and pepper. Stir a couple of times, then cover and simmer for 20 minutes until the rice is cooked and all the water absorbed. The rice should be fluffy and dry. Serve with a spring salad (see page *150*).

4 servings.

BEAN PIE

JAMAICA

8 oz (1¼ cups) white beans (soaked, see page 17)
1 tablespoon vinegar
2 teaspoons Worcestershire sauce
½ teaspoon mustard
2 oz (½ cup) grated cheese
salt and pepper to taste
2 teaspoons chopped mint
2 teaspoons chopped parsley

Cook the beans in plenty of water for about 1½ hours. Strain and retain the water. Add all the other ingredients, except the mint and parsley, to 2 tablespoons of the bean water, cook in a saucepan until the cheese melts. Toss the beans in this sauce, turn into a heatproof dish and grill until the top turns brown. Serve garnished with the mint and parsley.

4-6 servings.

PEA & ASPARAGUS BAKE

AUSTRALIA

1 lb asparagus
2 oz (4 tablespoons) butter
2 tablespoons flour
1 pint (2½ cups) milk
salt to taste
1 lb (3 cups) fresh green peas
1 oz (¼ cup) dry breadcrumbs
2 oz (½ cup) grated Cheddar cheese (mild)

Cut the asparagus into 1-inch pieces and steam until tender. Make a white sauce using half the butter, the flour and milk, seasoned with salt. Add the asparagus to the sauce and then add the uncooked peas. Oil a shallow pie dish and turn the vegetable and sauce mixture into it. Sprinkle with the dry breadcrumbs and grated cheese. Dot with the remaining butter and bake in the oven at 350°F, mark 4 until the sauce bubbles and the top is crusty and brown (about 45 minutes).

6-8 servings.

LIMA BEAN CASSEROLE

MEXICO

Milder than the usual Mexican dishes, this one allows the full flavour of the limas to come through.

12 oz (2 cups) cooked lima beans
¼ lb salt pork, cut in strips
1 large can tomatoes, chopped
1 large onion, finely chopped
1 tablespoon molasses
1 tablespoon sugar
¼ teaspoon chilli powder
salt to taste

Mix all the ingredients in a bowl and put them into a covered casserole. Cook in the oven at 350°F, mark 4 for 45 minutes. Serve with a bowl of grated cheese.

4 servings.

BROAD BEAN KUKU

IRAN

The most magnificent picnic I have ever enjoyed took place in the hills that surround Tehran. There were more people to carry the food than there were guests and each dish was served on a covered silver platter which had been wrapped in carpets to keep it warm. First temptation in this afternoon-long feast was this traditional cake-like omelette.

12 oz (2 cups) fresh broad beans
6 eggs
4 spring onions, chopped
1 teaspoon lemon juice
salt and pepper to taste
2 oz ($\frac{1}{4}$ cup) butter

Boil the beans until tender and drain. Beat the eggs and mix with the beans, onions, lemon juice, salt and pepper. Heat the butter in a frying pan until it is bubbling and cook the mixture, covered, for about 20 minutes over a low heat. Brown lightly under the grill. Serve hot or cold on a round platter and cut into cake-size wedges.

6 servings.

$1\frac{1}{2}$ lb (3 cups) minced (ground) beef
1 medium-sized onion, chopped
2 or 3 cloves garlic, peeled and crushed
1 tablespoon chopped parsley
2 teaspoons chopped fresh thyme
1 tablespoon minced green pepper
1 egg, beaten
2 oz (1 cup) fresh breadcrumbs
a dash of tabasco
salt and pepper to taste
3 oz ($\frac{3}{4}$ cup) chopped peanuts
4-5 lettuce leaves, shredded
3 tablespoons mayonnaise
6 soft bread rolls

Mix together all the ingredients except the peanuts, lettuce, mayonnaise and bread rolls. Allow to stand for 20 minutes for the flavours to mingle. Add the peanuts. Mix the shredded lettuce with the mayonnaise and warm the buns in the oven.

Shape the mixture into six patties and grill at high heat to stop the juice running out. Spread the mayonnaise mix over the burgers and serve in the buns.

6 servings.

ESAU'S POTAGE *RED LENTIL & RICE STEW*

MIDDLE EAST

This is the biblical potage for which Esau sold his heritage. Lentils are carried by the desert tribes in the Middle East because they are light in weight, don't perish on long journeys and can be made into all kinds of dishes. These are cooked in one huge pan when the tribe reaches a water hole.

5 medium-sized onions, chopped
4 oz (½ cup) butter
1½ lb (3 cups) red lentils
4 pints (10 cups) water
2 oz (¼ cup) natural rice
2 teaspoons salt

Fry the onions in the butter in a saucepan until golden, then add the lentils. Pour in the water, bring to the boil and simmer until it starts to thicken (about 30 minutes). Add the rice and salt and boil over a high heat for a couple of minutes and then reduce the heat and simmer, uncovered, until the rice is cooked – about 20 minutes. Serve with grilled pork chops or sausages.

6 servings.

2 oz ($\frac{1}{3}$ cup) muesli with raisins
4 tablespoons milk
1 egg, lightly beaten
6 oz (1 cup) cooked lima beans, puréed with a little liquid in a blender
2 tablespoons chopped parsley
$\frac{1}{4}$ teaspoon dried dill
4 oz ($\frac{1}{2}$ cup) cottage cheese
4 oz (1 cup) finely grated carrot
3 oz ($\frac{3}{4}$ cup) ground peanuts
2 tablespoons dry breadcrumbs
1 oz (2 tablespoons) butter
paprika

Soak the muesli in the milk. Add the egg and mix. Add all the other ingredients except the breadcrumbs, butter and paprika, and mix again. Press the mixture into a loaf tin, cover with the breadcrumbs and dot with the butter. Sprinkle the top with paprika to add colour. Bake in the oven at 350°F, mark 4 for 30 minutes. Serve hot or cold.

4 servings.

DHAL RICE

SRI LANKA

This is a basic dish in the villages of Sri Lanka where the houses are built of mud and thatch and where the food is eaten sitting cross-legged on the ground. It is also served as one of several dishes on the tables of Columbo.

1 lb (2 cups) small lentils
½ oz (1 tablespoon) fat
1 large onion, chopped
2 teaspoons curry powder
1-2 dry chillies
2-4 cardamom seeds, ground
1 lb (2 cups) rice
salt to taste

Soak the lentils for 20 minutes. Heat the fat in a saucepan and fry all the ingredients except the lentils and rice. When the onions are slightly brown add the rice and fry for a few minutes. Add water to ½ inch above the rice and bring to the boil. When the water has cooked away to the level of the rice, add the lentils and salt and cook carefully for about 20 minutes, until done. Serve in shallow bowls.

8 servings.

NEW ZEALAND

1 lb (2¼ cups) fresh green peas, mashed
1 egg
2 tablespoons vegetable oil
1 tablespoon diced onion
1 tablespoon lemon juice
1½ oz (½ cup) fresh wholewheat breadcrumbs
3 tablespoons cream
¼ pint (⅔ cup) vegetable stock
4 rashers of bacon, cooked and finely chopped

Mix all ingredients together. Stir thoroughly and put the mixture into an oiled casserole. Bake in the oven at 350°F, mark 4 for 45 minutes.

4 servings.

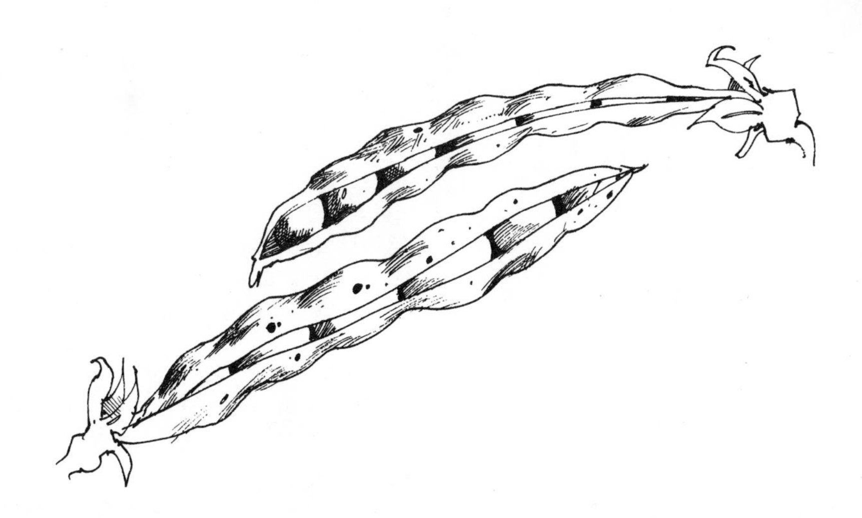

FLAGEOLETS A LA CREME

FRANCE

1 oz (2 tablespoons) butter
8 oz ($1\frac{1}{3}$ cups) cooked flageolets (see page 14)
4 oz ($\frac{1}{2}$ cup) double cream
1 teaspoon flour
salt to taste
2 sprigs of parsley, chopped

Melt the butter in a saucepan, add the flageolets and cook gently for 3 or 4 minutes. Mix the cream with the flour in a bowl and stir into the flageolets. Bring almost to the boil and season with salt.

Serve garnished with chopped parsley.

4 servings.

TUTU OF BLACK BEANS

BRAZIL

This is another Brazilian dish which shows the versatility of the shiny, tender black bean. It's a basic paste which is seasoned differently all around the country – at its hottest in Bahia where the descendants of Dahomey slaves use the red hot spices brought over by their ancestors.

1 lb (2 cups) black beans (soaked, see page 12)
4 tablespoons flour
salt and pepper to taste
1 bayleaf
1 medium-sized onion, sliced
2 tomatoes, sliced
2 tablespoons oil

Cook the beans in plenty of water for 1 hour. Drain them and purée in a blender, then put into a saucepan. Mix a little water into the flour and add to the bean purée, together with the salt, pepper and bayleaf. Heat gently, stirring all the time until the mixture looks like a paste. Remove the bayleaf and smooth into an ovenproof dish.

Sauté the onion and tomato slices carefully in the oil until tender and place them on top of the bean paste. (To make it hotter, add four finely chopped chilli peppers with the onion and tomatoes.)

Cook in the oven at 350°F, mark 4 for 25 minutes. Serve as a first course or as a side dish with grilled or roast pork.

6 servings.

GREEN BEANS & TOMATOES

GREECE

1 large onion, chopped
1 clove garlic, peeled and crushed
2 tablespoons oil
1 small can tomatoes
salt and pepper to taste
¼ pint (⅔ cup) water
1 lb French beans, shredded

In a saucepan sauté the onions and garlic in the oil until soft and golden. Add the tomatoes, salt and pepper and simmer for a few minutes. Pour in the water and bring to the boil. Add the beans and cook over a low heat until tender (about 30 minutes). Serve cold.

4 servings.

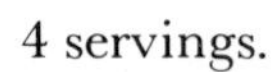

CARROT AND BEAN PUREE

CZECHOSLOVAKIA

8 oz (1 cup) cooked, mashed carrots
6 oz (1 cup) cooked, mashed white beans
salt and pepper to taste
¼ pint (⅔ cup) hot milk

Mix the carrots and beans together, season with salt and pepper and carefully add the hot milk. Blend again until creamy. Serve as an accompaniment to grilled pork chops or roast pork.

4 servings.

HARICOT A LA NEIGE

MARTINIQUE

1 oz (2 tablespoons) butter
8 oz (1¼ cups) cooked haricot or butter beans
2 tablespoons cream
salt and pepper to taste
1 egg, separated
1 tablespoon chopped parsley

Melt the butter in a saucepan, add the beans, cream, salt, pepper and the egg yolk. Cook gently for 3 minutes, stirring all the time. Fold in the stiffly-whisked egg white. Serve in a hot dish and garnish with parsley.

4 servings.

SWEET BEANS

SCANDINAVIA

12 oz (1¾ cups) haricot beans (soaked, see page 14)
2 tablespoons treacle or golden syrup
1 tablespoon vinegar
salt and pepper to taste

Boil the haricot beans in plenty of water for about 1 hour, or until soft. Drain and mix in the treacle or golden syrup, vinegar, salt and pepper. Serve hot.

6 servings.

PERIGORD PEAS

FRANCE

This recipe can be made with butter but in Perigord they use pork fat to make it richer and more substantial.

1 lb (2½ cups) small fresh peas
½ oz (1 tablespoon) butter or pork fat
1 heart of lettuce, cut in four
4 spring onions, chopped
2 rashers bacon, chopped
¼ pint (⅔ cups) water
salt and pepper to taste

Cook the peas, butter, lettuce, onions and bacon in the water for 10 minutes. Season with salt and pepper and cook until the peas are tender, about 5 minutes more. Drain and serve with grilled meats.
4 servings.

PEANUT BALLS I

½ oz (1 tablespoon) butter
2 tablespoons flour
4 tablespoons milk
½ teaspoon salt
pepper to taste
2 oz (½ cup) ground peanuts
10 oz (1¾ cups) cooked rice
1 egg, beaten

Melt the butter in a small saucepan, add the flour and cook until bubbling. Stir in the milk carefully to make a white sauce, Add the seasoning. Pour into a bowl and stir in the ground peanuts, rice and egg. Shape this mixture into small balls and sauté in an oiled pan until they are brown all over. Serve with a cheese sauce or just grated cheese sprinkled on top.

4 servings.

PEANUT BALLS II

6 oz ($1\frac{1}{2}$ cups) ground peanuts
6 oz ($1\frac{1}{2}$ cups) wholemeal breadcrumbs
1 teaspoon finely grated lemon peel
1 tablespoon chopped parsley
$\frac{1}{2}$ teaspoon celery salt
2 eggs

Mix the ground peanuts with the breadcrumbs, lemon peel, parsley and celery salt in a bowl. Beat the eggs and fold into the other ingredients. The mixture should be stiff but if too stiff, add a little vegetable stock. Roll into balls the size of a walnut and cook in a stew or thick soup for 35 minutes.

4 servings.

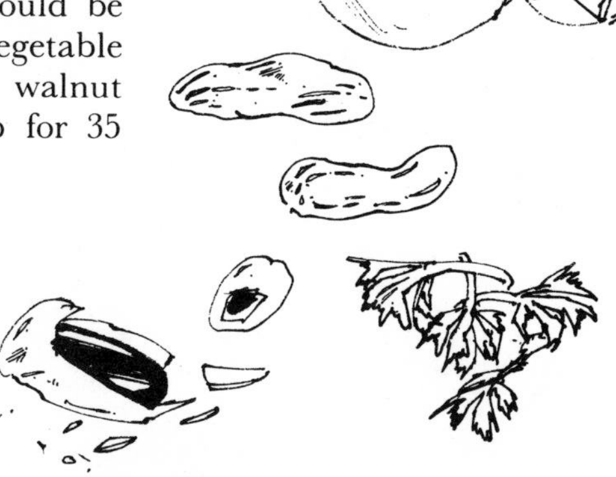

PARMESAN GREEN BEANS

FRANCE

1 lb fresh green beans
$\frac{1}{2}$ medium-sized onion, finely chopped
1 small clove garlic, peeled and crushed
3 tablespoons freshly grated Parmesan cheese
6 tablespoons olive oil
2 tablespoons white wine vinegar
salt and pepper to taste

Boil the beans for 2 minutes or until they are just tender. Drain. Mix with the onion, garlic, and Parmesan cheese. Pour over the oil and vinegar and season with salt and pepper. Chill.

Serve with a cold meat platter on hot summer days.

4 servings.

FRIED HARICOT BEANS

FRANCE

4 oz ($\frac{1}{2}$ cup) butter
8 oz ($1\frac{1}{3}$ cups) cooked haricot beans
(see page 14)
salt and pepper to taste

Heat the butter to smoking point and fry the beans until golden brown. Season with salt and pepper.

4 servings.

GREEK BEANS IN THE OVEN

GREECE

This recipe has the taste of ancient Greece in it, mixing fragrant honey and cloves with garlic and tomato.

2 medium-sized onions, chopped
3 tablespoons olive oil
1 clove garlic, peeled and crushed
2 tablespoons tomato paste
2 tablespoons vinegar
3 tablespoons honey
1 pint (2½ cups) hot water
2 cloves
1 lb (2¼ cups) cooked black-eyed beans (see page 12)
grated cheese, Kefaloteri or Parmesan

Sauté the onions in oil until soft. Add all other ingredients, except the beans and cheese and bring to the boil. In a shallow earthenware pot combine this mixture with the beans and bake for 30 minutes at 375° F, mark 5.

Serve with a bowlful of cheese to sprinkle on top. This is very good as the side-dish to roast lamb flavoured with oregano and thyme.

6 servings.

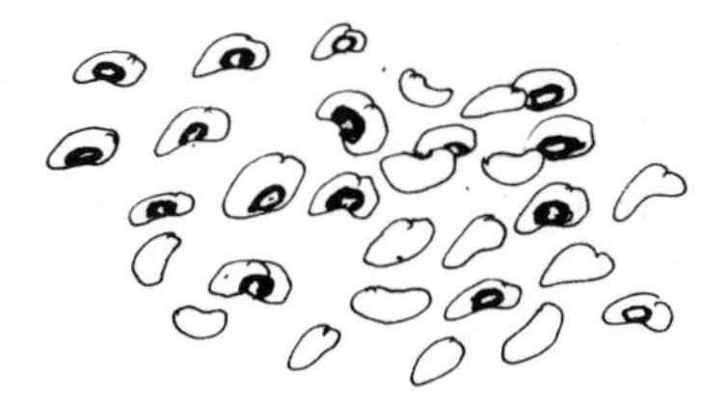

HARICOT BEANS BASQUE

FRANCE

3 tablespoons olive oil
1 clove garlic, crushed
8 oz (1⅓ cups) cooked haricot beans (see page 14)
1 tablespoon water
salt and pepper

Heat the oil in a frying pan, sauté the garlic until soft and add the haricot beans with the water. Season with salt and plenty of pepper. Serve steaming hot.

4 servings.

BROAD BEANS & ARTICHOKES

GREECE

2 lb (5 cups) broad beans
8 artichoke bottoms
1 tablespoon olive oil
½ teaspoon cornflour (cornstarch)
juice of 1 lemon
2 tablespoons chopped parsley

Cook the broad beans and artichoke bottoms separately. Strain and keep 2 tablespoons of the water in which the beans were cooked. Heat the oil in a saucepan, stir in the cornflour (cornstarch), bean water, lemon juice and parsley, simmer for 3 minutes more. Add the beans and artichokes and heat thoroughly, taking care not to soften the broad beans. Serve as a first course or with roast lamb.

6-8 servings.

LENTIL & RICE MOULD

NEW ZEALAND

This is a recipe from a book called Hints on Healthy Living written by my grandfather (Dr Ulric Williams) in New Zealand, long before the health movement, as we know it today, started.

His clinic and his advice were, to say the least controversial, treating, as he did, Maori and European alike by nature cure methods of his own devising. It is on his creed that I base many of the recipes for my family today.

8 oz (1 cup) lentils
6 oz (¾ cup) rice
1 large onion, chopped
1 teaspoon mixed herbs
1 oz (2 tablespoons) butter

Cook the lentils and rice separately, until tender. Once cooked, mix them together, add the other ingredients, steam in a buttered basin for 1½ hours and turn out when cold. Serve with salads.

4 servings.

ARTICHOKES WITH PEAS CARCIOFI CON PISELLI

ITALY

Traditionally, this dish is made with the soft, sweet, greenish gold piselli peas which are grown around Rome, but as their season is short and they are not readily available, any fresh, young, green peas will do just as well.

1 medium-sized can artichoke hearts
juice of 1 lemon
$1\frac{1}{2}$ oz (3 tablespoons) butter
2 tablespoons olive oil
$\frac{1}{4}$ lb prosciutto or lean bacon, chopped
1 very small onion, chopped
8 oz ($1\frac{1}{2}$ cups) fresh young peas (shelled)
a pinch of sugar
$\frac{1}{2}$ cup chicken stock
salt and pepper to taste

Drain the artichoke hearts, sprinkle them with the lemon juice and let them stand for 1 hour.

In a saucepan heat together the butter, olive oil and prosciutto. Add the artichoke hearts, peas, sugar, chicken stock, salt and pepper. Cover the saucepan and simmer gently for 20 minutes until tender. Serve garnished with sliced lemon, as a side dish with veal.

4 servings.

NEW ZEALAND

1 lb (2½ cups) fresh peas
1 oz (2 tablespoons) butter
1½ tablespoons flour
½ pint (1¼ cups) milk
salt and pepper to taste
3 tablespoons chopped fresh mint

Cook the peas and drain. In the meantime make a white sauce. Melt the butter in a small saucepan. Remove from the heat and stir in the flour. Once well blended, return the pan to the heat and gradually add the milk, stirring all the time. Bring to the boil and season to taste. Add the chopped mint and peas; reheat gently if necessary.

4 servings.

CREAMED SPLIT PEAS

CZECHOSLOVAKIA

8 oz (1 cup) split yellow peas
1½ oz (3 tablespoons) butter
1 teaspoon sugar
½ beef stock cube
1 tablespoon flour
4 tablespoons cold water
salt and pepper to taste

Cook the split peas in plenty of water until soft (about 30 minutes), and sieve. Melt the butter and sugar and stock cube in a saucepan, sprinkle in the flour and stir over low heat for 1 minute. Remove from the heat and stir in the cold water, salt and pepper. Add the sieved peas and cook gently in the sauce for 5 minutes.

4 servings.

BEAN DELIGHT

AUSTRIA

This dish is often served in Austrian skiing hotels – and very welcome it is too after all that exercise.

2 pints (5 cups) water
12 oz (1¾ cups) white beans (soaked, see page 17)
1 small onion, chopped
1 carrot, chopped
1 bayleaf
¼ teaspoon dried thyme
2 cloves
8 oz (1 cup) mashed potato
4 oz (½ cup) butter
salt to taste
¼ teaspoon sugar
¼ pint (⅔ cup) cream or milk

Bring the water to the boil and add the beans, onion, carrot, bayleaf, thyme and cloves. Lower the heat and simmer, covered, for 2 hours or until the beans are soft. Strain, mash (or purée in a blender) and mix with the mashed potato. Make a paste of the butter, salt and sugar and stir this into the purée. Quickly stir in the cream or milk and serve steaming hot in a hot dish.

4 servings.

ITALIAN PEAS

ITALY

1 tablespoon olive oil
8 oz (1 cup) cooked ham, cubed
1 tablespoon finely chopped onion
1 teaspoon finely chopped parsley
½ teaspoon finely chopped basil
¼ pint (⅔ cup) water
a large pinch of pepper
1 lb (2½ cups) fresh peas
½ oz (1 tablespoon) butter
salt to taste

Heat the oil in a frying pan and sauté the ham, onion, parsley and basil over a low heat until the onion is soft and transparent. Add the water and pepper and stir. Add the peas and butter and simmer gently, covered, for 10-15 minutes. Add salt to taste.

4 servings.

BEANS IN WINE

FRANCE

1 medium-sized onion, chopped
1 oz (2 tablespoons) butter
6 oz (¾ cup) smoked ham, diced
salt and pepper to taste
12 oz (2 cups) cooked red kidney beans
1 pint (2 ½ cups) red wine

Sauté the onion in the butter until soft. Add the smoked ham and fry until brown all over. Season with salt and pepper.

Put this mixture into a casserole with the beans and wine, stir and cook, uncovered, in the oven at 350°F, mark 4 for 30 minutes. Serve with grilled steaks or on its own with a green salad with a lemon dressing and garlic bread.

4 servings.

BOHENSCHNITZEL *BEAN RISSOLES*

AUSTRIA

1 lb (2 cups) white beans (soaked overnight)
½ onion, chopped
2 eggs
4 sprigs parsley, chopped
salt to taste
8 oz (1 cup) diced ham
1 tablespoon flour
2 oz (1 cup) fresh breadcrumbs

Cook the beans in plenty of water for 1½ hours. Drain and purée in a blender (or put through a sieve). Fry the onions and add to the pureé with 1 egg, the parsley, salt and ham. Mould the mixture into patties or rissole shapes. Beat the second egg. Mix the flour with the breadcrumbs and dip each rissole first into the beaten egg, then into the breadcrumbs mixture and fry in hot fat.

These are delicious served with a hot cauliflower salad (see page *149*).

4 servings.

SWISS BEANS

SWITZERLAND

This is gnome food for the kind of gnomes that are found in and around the banking centres, not amongst the wild flowers on Swiss mountains.

12 oz (2 cups) cooked butter beans
4 oz (½ cup) butter
2 egg yolks, beaten
½ pint (1¼ cups) milk
salt to taste

Mix the beans, butter, egg yolks, milk and salt in a bowl. Turn into a casserole and bake, uncovered, in the oven at 350°F, mark 4 for 30 minutes or until the top is brown. Serve with veal escalope or chops.

6 servings.

PEAS & MUSHROOMS PISELLI AL FUNGHI

ITALY

Just one of the recipes from the area around Naples where beans and peas are a basic part of the diet and ways of cooking them are original and always changing.

8 oz (2 cups mushrooms, finely chopped
1 small carrot, sliced
1 lettuce heart, shredded
8 oz (1¼ cups) peas (freshly shelled if possible)
a sprig of parsley
¼ teaspoon thyme
1 oz (2 tablespoons) butter
salt and pepper to taste
1 teaspoon sugar
3 tablespoons water

Put all the ingredients into a heavy saucepan. Cover the pan and put it on an asbestos mat over low heat for about 45 minutes stirring from time to time, until the peas are tender and the liquid almost entirely evaporated. Serve with veal or fish.

4 servings.

ITALY

The Italians have a tradition of eating peas which stretches back to the ancient Romans, although judging from Apicius' cookbook, they used dried ones. In Piedmont, where this recipe comes from, a love of spices even influences the way vegetables are cooked, as seen with the nutmeg, added here.

1 lb (2½ cups) shelled peas
1 head of fennel
1½ oz (2½ tablespoons) butter
a dash of nutmeg
salt and pepper to taste

Boil the peas in salted water with the fennel until tender. Drain, purée the fennel in a blender (or chop and rub through a sieve). In a wide saucepan, melt the butter, add the peas, nutmeg and salt and pepper. Stir well, add the fennel purée, and heat gently for 2 or 3 minutes. Serve with grilled fish.

4 servings.

LENTIL SALAD

AUSTRIA

12 oz (1½ cups) continental lentils
½ onion, finely chopped
4 tablespoons olive oil
4 tablespoons wine vinegar
½ teaspoon cumin
½ teaspoon French mustard
salt and pepper to taste

Cook the lentils in plenty of water for 30 minutes or until tender. Drain and mix in the onion. Mix together the oil, vinegar, cumin, mustard, salt and pepper and pour over the lentils. Toss, cover and refrigerate for 2 hours.

6 servings.

This salad is guaranteed to revive even the most exhausted member of the family. In my home we eat it at least once a week.

4 oz (¾ cup) cooked soy beans (see page 16)
4 oz (1 cup) chopped peanuts
1 stalk of celery, chopped
½ small onion, finely chopped
½ green or red pepper, chopped
4 oz (¾ cup) sultanas
4 oz (1½ cups) sprouted mung beans (see page 00)
2 tablespoons parsley, roughly chopped
1 tablespoon kelp (granulated)
1 tablespoon sesame seeds
4 oz (1½ cups) cabbage, finely chopped

For the dressing:
6 tablespoons sunflower oil
2 tablespoons fresh lemon juice
sea salt
1 clove garlic, peeled and crushed

Mix all the salad ingredients together and vary the taste each time by adding chopped apple, tomato or fresh pineapple. Combine the ingredients for the dressing, blend thoroughly and pour over the salad and leave to stand for 15 minutes before serving with wholemeal bread as a complete meal.

4 servings.

THREE BEAN SALAD

GREECE

Plato thought that eating beans was one of the important ways to a long life. Certainly there were on the distinguished tables of his times, many bean salads which were appreciated also for their colour and different tastes.

An attractive addition to any summer meal is the Three Bean Salad. I have chosen contrasting beans but it also looks (and tastes) marvellous with all white and cream beans, or shades of greens, browns and reds.

5 oz ($\frac{3}{4}$ cup) black-eyed beans (soaked, see page 12)
5 oz ($\frac{1}{2}$ cup) green flageolets (soaked, see page 14)
5 oz ($\frac{1}{2}$ cup) red kidney beans (soaked, see page 14)
olive oil
white wine vinegar
1 onion, thinly sliced
a large bunch of parsley, chopped
salt and pepper to taste

Cook the three kinds of soaked beans separately or the tastes will mingle – also the black-eyed beans and flageolets will take about half as long to cook as the kidney beans. When they are all tender, rinse and drain together.

For the dressing, mix the oil and vinegar in the ratio of 3 to 1, then add the onion, parsley and plenty of salt and pepper. Pour over the beans well in advance of serving to allow it to soak in. Do not chill, as the beans have more flavour when still warm.

The salad looks stunning served in a shallow white bowl.

4 servings.

See colour picture

BEAN SPROUT SALAD I

This recipe combines Chinese and western ingredients to make a delicious, crunchy and healthy salad.

12 oz (3 cups) sprouted mung beans
1 lb Chinese cabbage, finely chopped
½ onion, finely chopped
2 oz (½ cup) cashew nuts, chopped
4 oz (¾ cup) Edam cheese, cubed

For the dressing:
2 tablespoons lemon juice
6 tablespoons oil
1 clove garlic, peeled and crushed
salt and freshly ground black pepper

Mix all the salad ingredients together. Combine the lemon juice, oil, garlic, salt and pepper and blend thoroughly. Pour over the salad at the last moment before serving with an accompanying bowl of yoghurt.

6 servings.

RAW PEA SALAD

AUSTRALIA

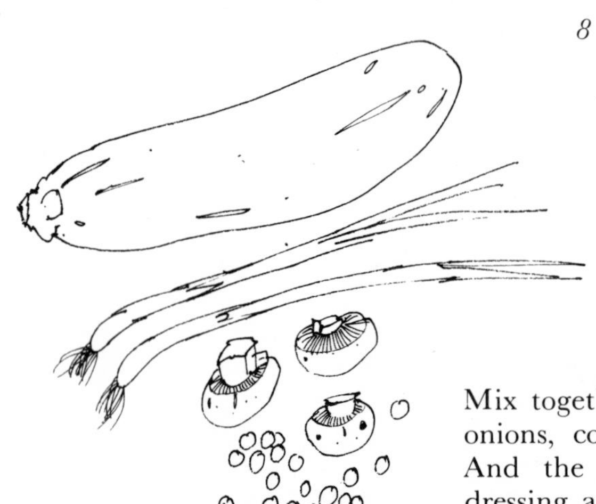

8 oz ($1\frac{1}{4}$ cups) raw shelled peas (young and tender)
4 oz (1 cup) small button mushrooms, thinly sliced
2 spring onions, thinly sliced
1 small raw courgette, cubed
a sprig of fresh basil, chopped
salt and pepper to taste
1 teaspoon sweet or semi-sweet sherry
oil and lemon dressing (see page 150)

Mix together the peas, mushrooms, spring onions, courgette, basil, salt and pepper. And the sherry to the oil and lemon dressing and stir into the salad. Serve the salad in small letuce leaves.

2 servings.

RED BEAN & DAMSON SALAD

RUSSIA

$1\frac{1}{2}$ tablespoons damson jam
$\frac{1}{2}$ teaspoon red wine vinegar
1 small clove garlic, peeled and crushed
salt to taste
$\frac{1}{2}$ teaspoon basil
$\frac{1}{4}$ teaspoon coriander
8 oz ($1\frac{1}{3}$ cups) cooked red beans

Sieve the damson jam or pureé in a blender and heat in a small saucepan with the vinegar. Add the garlic, salt, basil and coriander and simmer carefully for 2 minutes. Stir this mixture into the beans, cover and leave out for 2 hours so that the beans can absorb the flavours. Serve chilled.

4 servings.

BEAN SPROUT SALAD I I

CHINA

8 oz (2 cups) sprouted mung beans
6 oz (1½ cups) cubed apples
1 medium-sized cooked beetroot, cubed
1½ oz (⅓ cup) chopped walnuts
a sprig of fresh mint, chopped
1 oz (2 tablespoons) sultanas (soak and drain if hard)
juice of ½ lemon
6 tablespoons (½ cup) tahini
about 3 tablespoons (¼ cup) fresh orange juice

Mix all the salad ingredients with the lemon juice. Dilute the tahini with the orange juice to desired consistency and fold into the salad just before serving.

4 servings.

LAYER SALAD

I grew up on various versions of my mother's Layer Salads and when she brought them to the table in a big glass bowl with all the colours of the countryside in full view, they never failed to draw whoops of appreciation. On the practical front, they are a marvellous way to empty the fridge but it is important to balance the sweet tastes with the vegetables, nuts, eggs, cheese, tuna or cold meats.

6 oz (1 cup) cooked soya beans
4 oz (1 cup) finely grated celery
6 oz (2 cups) shredded cabbage
6 oz (1½ cups) finely grated carrot
4 oz (1 cup) finely grated raw beetroot
3 oz (¾ cup) chopped peanuts
2 apples, grated
2 hard-boiled eggs, coarsely grated
8-10 chopped dates
2 oz (1½ cups) finely chopped parsley
4 oz (1 cup) grated mild cheese

Optional variations:
1 tin tuna, cubed ham, salami
sliced bananas
sliced tomatoes

In a large bowl, preferably glass as explained above, layer the ingredients, finishing with a layer of parsley. Layers of tinned tuna, cubed ham, salami or any other cold meat can be added. Sliced tomato or banano looks good laid on top of the parsley but do not mix in as they get mushy.

This versatile salad can be used as a complete meal if served with a buttered baked potato in its jacket.

Serve with a choice of dressings in large quantities; oil and lemon, yoghurt, tahini thinned with fresh orange juice and a home made mayonnaise.

4 servings.

FAVA BEAN SALAD

LEBANON

1 lb (2½ cups) broad beans
1 clove garlic
salt to taste
1 pod hot pepper
2 tablespoons olive oil
juice of ½ lemon
a sprig of parsley, chopped

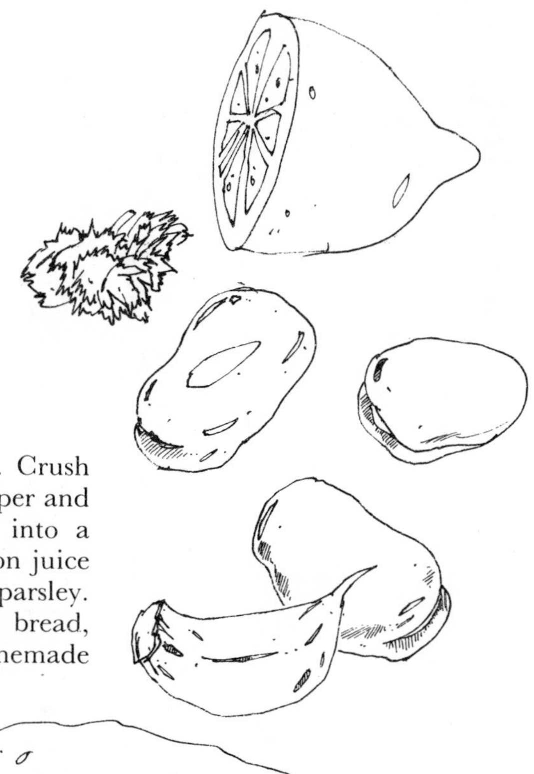

Boil the broad beans until tender. Crush the garlic with the salt and hot pepper and mash with the broad beans. Put into a shallow dish, pour the oil and lemon juice over and garnish with chopped parsley. Serve with kebabs and flat *pitta* bread, bought from Greek stores, or homemade (see page *157*).

4 servings.

BUTTER BEAN SALAD

AUSTRIA

8 oz (1½ cups) cooked butter beans
salt and pepper to taste
1 teaspoon chopped onion
1 teaspoon chopped parsley
2 tablespoons oil
juice of 1 lemon

Heat the beans in a little salted water. When the liquid is absorbed remove from the heat and mix with the pepper, onion, parsley, oil and lemon. Chill well.

4 servings.

BEAN SPROUTS

Soya beans, mung beans and also alfalfa seeds sprout very easily and make delicious, vitamin-rich additions to salads.

Soak 3 tablespoons of the beans or seeds overnight. Remove any that float to the surface as these won't be fertile and therefore won't sprout. Drain the rest and put in a wide-topped jar, fill with water and leave in a dark place. Rinse and drain three times a day and within 3-5 days the sprouts will be ready for eating.

Soya beans are best eaten when ½-1 inch long, mung beans 1-2½ inches and alfalfa about 1 inch. They will keep for 3-5 days in the fridge in a covered container.

Enjoy them raw, tossed in salads or in

sandwiches, or sauté some chopped onion and sweet pepper with the sprouts and season with soy sauce.

MARINATED BEAN SALAD

AMERICA

A recipe that owes its origins to both sides of the Californian/Mexican border.

4 oz ($\frac{2}{3}$ cup) cooked chick peas (see page 13)
8 oz green string beans, cooked and cut into 1 inch pieces
4 oz ($\frac{2}{3}$ cup) cooked red kidney beans (see page 14)
1 large onion, sliced into rings
1 pimiento, chopped
6 tablespoons olive oil
6 tablespoons wine vinegar
1$\frac{1}{2}$ tablespoons sugar
salt and pepper to taste

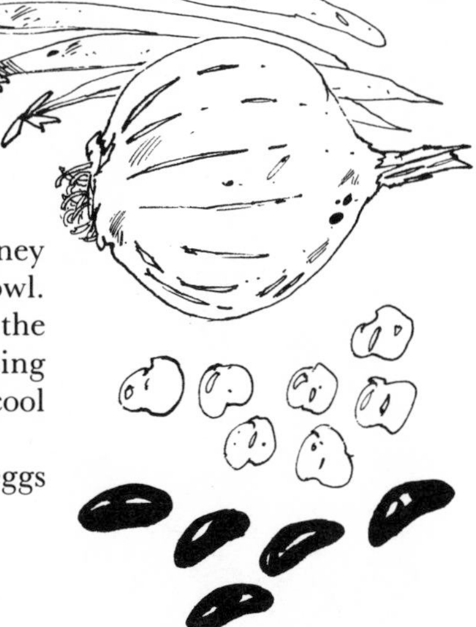

Put the chick peas, green beans, kidney beans, onion and pimiento into a bowl. Mix the rest of the ingredients to make the dressing and pour over the beans, tossing carefully. Cover the bowl and put in a cool place to marinate overnight.

This salad, served with hard-boiled eggs is good picnic food.

4 servings.

CHICK PEA SALAD

MIDDLE EAST

a generous handful of flat-leaf parsley, chopped
1 medium-sized onion, finely chopped
1 clove garlic, peeled and crushed
3 tablespoons olive oil
1 tablespoon yoghurt
2 tablespoons fresh lemon juice
salt and cayenne pepper to taste
8 oz (1⅓ cups) cooked chick peas (see page 13)

Put all the ingredients (except the chick peas) into a screw-topped jar and shake vigorously until the mixture looks creamy. Pour this dressing over the chick peas in a bowl, stir well and serve immediately – otherwise the chick peas will lose their crunchiness.
4 servings.

SOYA BEAN & POTATO SALAD

This is another of my inventive mother's marvellous salads.

6 oz (1 cup) cooked soya beans (see page 17)
12 cooked potatoes, diced
8 oz (2 cups) raw button mushrooms, sliced
2 oz (½ cup) finely chopped celery
2 spring onions, finely chopped (including green stalks)
8 green olives, chopped
salt and pepper to taste
3 tablespoons yoghurt or mayonnaise

Mix together all the ingredients except the yoghurt or mayonnaise and leave, covered, for 2 hours. Fold in the yoghurt or mayonnaise just before serving. Serve with cold meats or barbecues; good for buffets.

4-6 servings.

BEAN SALAD WITH EGGS

TURKEY

Stop at any small café in a small Turkish village and you will be invited into the kitchen to choose your meal from a choice of two or three dishes of the day, simmering juicily on the stove. This salad often arrives as an unpaid-for extra.

8 oz (1 cup) haricot beans (soaked, see page 14)
4 tablespoons olive oil
juice of 1 lemon
salt and pepper to taste
1 tomato, sliced
2 hard-boiled eggs, quartered
8 black olives, halved and pitted

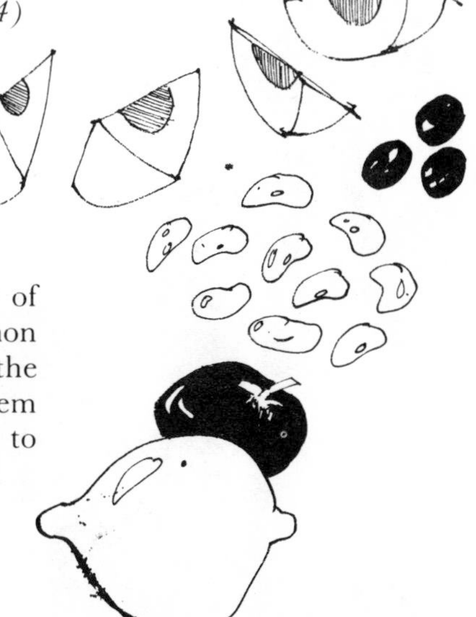

Cook the beans for two hours in plenty of water and drain. Add the olive oil, lemon juice, salt and pepper right away. Add the tomato, eggs and olives and mix them gently into the beans, taking care not to break up any of the ingredients.

4 servings.

12 oz (2 cups) cooked kidney beans, still warm
3 tablespoons oil
1 tablespoon tarragon vinegar
1 large firm tomato, chopped
½ small onion, finely chopped
1 or 2 canned pimientos
½ green pepper, finely chopped
a sprig of fresh sweet basil, finely chopped

While the kidney beans are still warm, but not hot, stir in the oil and tarragon vinegar and allow to stand until just before serving. Add the remainder of the ingredients and toss together lightly.

As an alternative, fresh or tinned pineapple pieces and chopped peanuts may be added instead of the tomato and onion.

4 servings.

HOT CAULIFLOWER SALAD

1 cauliflower, trimmed
2 tablespoons vinegar
6 tablespoons olive or salad oil
$\frac{3}{4}$ teaspoon salt
freshly ground black pepper
1 tablespoon chopped capers

Cook the whole cauliflower in boiling salted water until just tender. Blend together the vinegar, oil, salt and pepper and pour over the hot, drained cauliflower. Sprinkle the chopped capers on top and serve immediately.

This salad goes particularly well with Bean Rissoles (see page *132*).

4 servings.

SPRING SALAD

This salad is unusual both in its appearance and its taste.

6 oz (2 cups) shredded raw cabbage
4 oz (1 cup) chopped raw spinach
2 oz (½ cup) finely chopped onion
1 bunch of watercress, chopped
4 oz (1 cup) grated cheese
4 oz (1 cup) grated carrot
¼ teaspoon salt
freshly ground black pepper
1 tablespoon vinegar
2 tablespoons olive or salad oil

Toss together the cabbage, spinach, onion and watercress. Mix the cheese and carrot together and form into small balls the size of walnuts.

Make the dressing by mixing the salt and pepper with the vinegar and then whisking in the oil. Pour the dressing over the salad and top with the cheese and carrot balls.

Serve as an accompaniment to Congris (see page *109*).

4 servings.

LEMON DRESSING FOR GREEN SALAD

For the dressing:

2-3 tablespoons olive or salad oil
salt and pepper
1 tablespoon lemon juice

Mix the oil with the salt and pepper, then whisk in the lemon juice with a fork.

In a salad bowl combine two or more green salad ingredients such as lettuce, endive, watercress, chicory, celery, cress, cucumber, chopped fresh parsley, basil, chives. Add the dressing just before serving and toss lightly.

CUCUMBER AND YOGHURT SALAD

This is a delicious and refreshing salad which goes well with many of the Middle Eastern dishes.

2 medium-sized cucumbers, peeled and diced
salt
2 cloves garlic, peeled and crushed
¾ pint (2 cups) plain yoghurt
3 tablespoons chopped fresh mint
freshly ground black pepper

Sprinkle the cucumber with salt and leave to drain for 30 minutes. Add the garlic to the yoghurt and mix well. Stir in the mint and plenty of pepper. Add more salt if necessary. Drain the cucumber and mix into the yoghurt.

Chill and serve garnished with a sprinkling of chopped mint.

6 servings.

The proportion of oil to vinegar or lemon juice can vary according to individual taste – in the Middle East they prefer lemon juice to vinegar as they like sharp dressings. Sometimes they will use equal amounts of lemon juice and oil, and it really depends on the cook as to what she thinks is best for the particular salad.

Here are the basic proportions to work from:

1 tablespoon wine vinegar or lemon juice
3 tablespoons olive oil
1 clove garlic, peeled and crushed
salt and freshly ground black papper

Combine all the ingredients thoroughly together before pouring over the salad at the last minute before serving – unless it is a bean salad which needs to soak up the dressing.

Chopped parsley or finely chopped onion can be used as alternatives to the garlic.

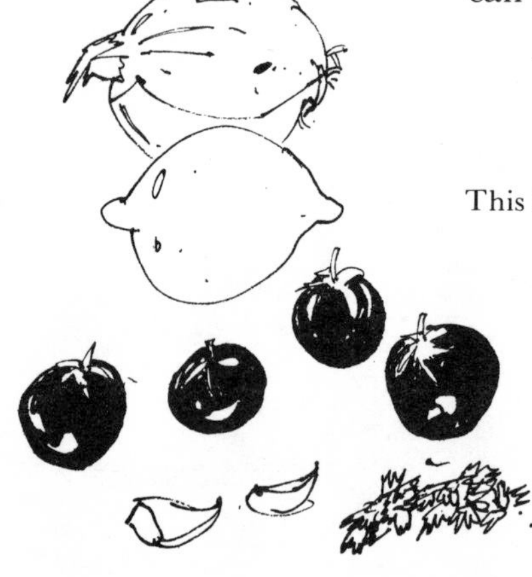

TOMATO & ONION SALAD

This is an attractive salad which is very good served with bean salads.

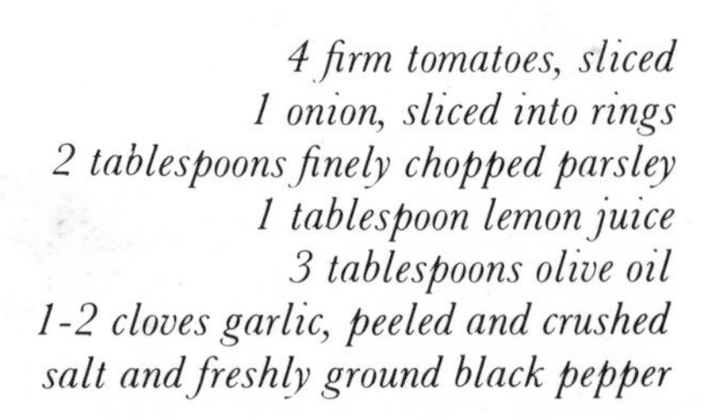

4 firm tomatoes, sliced
1 onion, sliced into rings
2 tablespoons finely chopped parsley
1 tablespoon lemon juice
3 tablespoons olive oil
1-2 cloves garlic, peeled and crushed
salt and freshly ground black pepper

Arrange the tomatoes and onions in overlapping rows on a serving plate. Sprinkle with the chopped parsley. Combine the lemon juice, olive oil, garlic, salt and pepper. Mix thoroughly and pour over the salad.

4 servings.

SALATA BALADI *PEASANT SALAD*

This tasty, fresh salad goes very well with many of the bean meals but is particularly good with Ful Medames (see page 75).

½ lettuce, finely shredded
8 oz (2 cups) chopped tomatoes
8 oz (2 cups) diced cucumber
4 oz (1 cup) chopped onions
1 tablespoon chopped mint
¼ teaspoon salt
freshly ground black pepper
1 tablespoon vinegar
3 tablespoons olive or salad oil

Toss together the lettuce, tomatoes, cucumber, onions and mint. Make the dressing by mixing the salt and pepper with the vinegar. When well blended, beat in the oil gradually with a fork and pour over the salad just before serving. If the dressing is left to stand, the oil will separate out, so if necessary whisk again just before using.

4 servings.

SOFT CREAM CHEESE LABNE

LEBANON

1 pint (2½ cups) plain yoghurt
½-1 tsp salt

Pour the yoghurt into a sieve lined with butter muslin or a fine cotton cloth. Let it stand overnight until the whey has drained away and a soft creamy curd is left. Season the curd to taste and serve as part of a *mezze* (mixed hors d'oeuvre) with Hummus bi Tahini (see page *40*) and Arab bread (see page *157*).

YOGHURT DRESSING

A light, fresh dressing for bean and lentil salads.

6 fl oz (1 cup) plain yoghurt
1 tablespoon fresh lemon juice
¼ teaspoon finely grated lemon rind
¼ teaspoon cumin
sea salt and freshly ground pepper to taste

Shake all the ingredients in closed jar and pour over at the last minute. Chopped fresh mint can be used as an alternative to cumin.

This is a delicious combination of kidney beans, fresh vegetables, vinegar and spices which makes a tasty addition to many a meal.

1 lb French beans, chopped
2 medium-sized onions, chopped
4 red or green peppers, seeded and chopped
12 oz (2 cups) cooked kidney beans
8 oz (1½ cups) tinned sweetcorn, drained (or fresh, cooked)
1 pint (2½ cups) cider or wine vinegar
6 oz (⅔ cup, firmly packed) sugar, preferably soft brown
2 tablespoons dry mustard
3 tablespoons mustard seed
2 teaspoons turmeric

Cook the French beans, onions and peppers in boiling salted water for about 8 minutes until barely tender. Drain, and combine with the kidney beans and sweetcorn.

In a saucepan, heat together the vinegar, sugar, mustard, mustard seed and turmeric. Cook gently until the sugar has dissolved, then bring to the boil and boil for 2 minutes. Add the vegetables and simmer gently for a further 5 minutes.

Pack into hot sterilised jars and seal at once.

Yields about 4 lb

1 oz (1 cake compressed) yeast,
(or 1 tablespoon dried)
¾ pint (2 cups) tepid water
1 ¼ lb (4 cups) wholewheat flour
4 oz (1 cup) plain white flour
4 oz (1 cup) cornmeal
2 teaspoons salt
1 tablespoon molasses or honey
2 tablespoons melted butter

Blend the fresh yeast with a little of the water. (For dried yeast, mix with 1 teaspoon sugar and then stir into the tepid water and leave until it froths.)

Mix together the flours, cornmeal and salt. Combine the yeast mixture with the rest of the water and the molasses or honey and stir this into the dry mixture, adding sufficient melted butter to make a firm dough. Add more white flour or water as necessary.

Turn the dough on to a floured board and knead until smooth and elastic. Put into a greased bowl, cover with a cloth and leave in a warm draught-free place (75-80°F) until doubled in size. Knead again and shape into 2 loaves and put in 2 greased and floured loaf tins. Brush with oil, cover with a cloth and leave to rise until the dough rises to the tops of the tins.

Bake in the oven at 400°F, mark 6 for 15 minutes, then reduce the temperature to 375°F, mark 5 and cook for about a further 35 minutes until the loaves sound hollow when tapped on the bottom.

PITTA BREAD

MIDDLE EAST

This is the traditional flat Greek or Arab bread that is the ideal accompaniment to so many meals. It can be bought fresh in Greek or Cypriot stores but is particularly delicious homemade – and very little trouble to make while waiting for beans to cook.

½ oz fresh (½ cake compressed) yeast
½ pint (1¼ cups) water
1 lb (4 cups) plain flour
½ teaspoon salt
oil

Blend the yeast in a little of the water until smooth, then add the rest of the water. Sift the flour and salt into a bowl and make a well in the centre. Pour in the yeast mixture and stir, gradually incorporating all the flour, until the dough is firm. Knead until it is smooth and elastic – about 10 minutes.

Brush oil all over the ball of dough to prevent the surface drying out. Place it in an oiled bowl, cover with a cloth and leave in a warm place for about 2 hours until the dough has doubled in size. Knead again then divide the dough into 8. Knead each into a ball, then flatten to about ¼ inch thick. Place on a flat, floured tray, cover with a cloth and leave to rise until spongy.

Preheat the oven to 450°F, mark 8 and heat an oiled baking tray. Lift the dough carefully on to the tray, brush with cold water and bake for 10 minutes near the top of the oven. Cool the bread on a wire rack. The bread should be pale, with a soft inside.

List of Suppliers

Dried beans and speciality foods are now widely available – in local Greek, Italian or Indian stores as well as in supermarkets and general delicatessens, but here is a list of a few of the many suppliers in the UK and USA.

Ceres
269 Portobello Road
London W 11
tel: 01-299 5571

T. Spencer
61 Lancaster Road
London W 11
tel: 01-727 4412

Cranks Health Food Shop
8 Marshall Street
London W 1
tel: 01-437 2915

I. Camisa & Son
61 Old Compton Street
London W 1
tel: 01-437 4686

Centro Espanol
74-76 Old Compton Street
London W 1

Loon Fung Chinese
Supermarket
39 Gerrard Street
London W 1
tel: 01-437 1922

Better Foods Foundation
300 North Washington St
Greencastle Pa 17225

Dutch School
22 North 7th St
Akron, Pa. 17501

Lekvar-by-the-Barrel
1577 First Avenue
New York 10028

Natural Foods Distributors
519 Monroe St
Toledo, Ohio 43604

Sahadi Importing Co
187 Atlantic Avenue
Brooklyn NY 11201

Vita Green Farms
P.O. Box 878
Vista, Ca 92803

INDEX